MOLECULAR BIOLOGY
INTELLIGENCE
UNIT

T0203722

Planaria:
A Model for Drug Action and Abuse

Robert B. Raffa, PhD
Department of Pharmaceutical Sciences
Temple University School of Pharmacy
Philadelphia, Pennsylvania, USA

Scott M. Rawls, PhD
Department of Pharmaceutical Sciences
Temple University School of Pharmacy
Philadelphia, Pennsylvania, USA

CRC Press
Taylor & Francis Group
Boca Raton London New York

CRC Press is an imprint of the
Taylor & Francis Group, an **informa** business

PLANARIA:
A MODEL FOR DRUG ACTION AND ABUSE
Molecular Biology Intelligence Unit

First published 2008 by Landes Bioscience

Published 2018 by CRC Press
Taylor & Francis Group
6000 Broken Sound Parkway NW, Suite 300
Boca Raton, FL 33487-2742

First issued in paperback 2019

No claim to original U.S. Government works

ISBN 13: 978-0-367-44608-6 (pbk)
ISBN 13: 978-1-58706-332-9 (hbk)

Visit the Taylor & Francis Web site at
http://www.taylorandfrancis.com

and the CRC Press Web site at
http://www.crcpress.com

Library of Congress Cataloging-in-Publication Data

Planaria : a model for drug action and abuse / [edited by] Robert B. Raffa, Scott M. Rawls.
   p. ; cm. -- (Molecular biology intelligence unit)
Includes bibliographical references and index.
ISBN 978-1-58706-332-9
1. Planaria--Effect of drugs on. 2. Planaria as laboratory animals. 3. Pharmacology, Experimental. 4. Drug abuse--Research--Methodology. I. Raffa, Robert B. II. Rawls, Scott M. III. Series: Molecular biology intelligence unit (Unnumbered)
[DNLM: 1. Planarians--drug effects. 2. Models, Animal. 3. Planarians--physiology. 4. Substance-Related Disorders. QX 352 P699 2008]
RM301.25.P63 2008
615'.78--dc22

2008045513

Dedication

To our families and our mentors, with love and gratitude

About the Editors...

ROBERT RAFFA is Professor of Pharmacology and Chair of the Department of Pharmaceutical Sciences at Temple University School of Pharmacy in Philadelphia, Pennsylvania. He has bachelor's degrees in Chemical Engineering and Physiological Psychology, master's degrees in Biomedical Engineering and Toxicology, and a doctorate in Pharmacology. He is the co-author or editor of several books on pharmacology, has published over 200 articles in refereed journals, and is active in editorial and professional society activities. His laboratory conducts NIH-funded research on drug action and abuse using the planarian model.

About the Editors...

SCOTT M. RAWLS is an Assistant Professor of Pharmacology in the Department of Pharmaceutical Sciences at Temple University School of Pharmacy and Research Professor at the Center for Substance Abuse Research in Philadelphia, Pennsylvania. Dr. Rawls holds a bachelor's of science degree in Chemistry and doctorate in neuroscience.

He maintains an active research laboratory funded by the National Institutes of Health (NIH) and the Pennsylvania Department of Health and has published over 40 articles in peer-reviewed journals since 1998. Dr. Rawls is active in the education of professional and graduate students. In the professional curriculum, Dr. Rawls coordinates the Anatomy and Physiology course and teaches the cardiovascular and renal sections in the Pharmacology course. He is the recipient of several teaching awards, including the American Association of Colleges of Pharmacy Teaching Award in 2006 and Biological Sciences Teaching Award at Washington College in 2003. Dr. Rawls is also the co-author of a pharmacology text, *Netter's Illustrated Pharmacology*.

CONTENTS

EDITORS

Robert B. Raffa
Department of Pharmaceutical Sciences
Temple University School of Pharmacy
Philadelphia, Pennsylvania, USA
Email: robert.raffa@temple.edu
Chapters 1, 8, 9

Scott M. Rawls
Department of Pharmaceutical Sciences
Temple University School of Pharmacy
Philadelphia, Pennsylvania, USA
Email: srawls@temple.edu
Chapter 7

CONTRIBUTORS

Note: Email addresses are provided for the corresponding authors of each chapter.

Charles I. Abramson
Department of Psychology
Oklahoma State University
Stillwater, Oklahoma, USA
Chapter 11

Kiyokazu Agata
Department of Biophysics
Graduate School of Science
Kyoto University
Kitashirakawa-Oiwake
Sakyo-ku, Kyoto, Japan
Email: agata@mdb.biophys.kyoto-u.ac.jp
Chapter 2

Antonio Carolei
Department of Neurology
University of L'Aquila
L'Aquila, Italy
Chapters 4, 5

Irene Ciancarelli
Department of Neurology
University of L'Aquila
L'Aquila, Italy
Chapter 4

Marco Colasanti
Dipartimento di Biologia
Università di Roma Tre
Viale Marconi
Roma, Italy
Email: colasant@uniroma3.it
Chapters 5, 6

Takashi Gojobori
Center for Information Biology and DNA
    Databank for Japan
National Institute of Genetics
Yata, Mishima
Shizuoka, Japan
Email: tgojobor@genes.nig.ac.jp
Chapter 3

Kazuho Ikeo
Center for Information Biology and DNA
    Databank for Japan
National Institute of Genetics
Yata, Mishima
Shizuoka, Japan
Chapter 3

Yoshihisa Kitamura
Department of Neurobiology
Kyoto Pharmaceutical University
Misasagi, Yamashina-ku
Kyoto, Japan
Email: yo-kita@mb.kyoto-phu.ac.jp
Chapter 2

Michael Levin
Center for Regenerative
 and Developmental Biology
Forsyth Institute
and
Department of Developmental Biology
Harvard School of Dental Medicine
Boston, Massachusetts USA
Email: mlevin@forsyth.org
Chapters 11, 12

Katsuhiko Mineta
Graduate School of Information Science
 and Technology
Hokkaido University
Sapporo, Hokkaido, Japan
Chapter 3

Cindy L. Nicolas
Center for Regenerative
 and Developmental Biology
Forsyth Institute
and
Department of Developmental Biology
Harvard School of Dental Medicine
Boston, Massachusetts USA
Email: cnicolas@forsyth.org
Chapter 11

Kaneyasu Nishimura
Department of Biophysics
Graduate School of Science
Kyoto University
Kitashirakawa-Oiwake
Sakyo-ku, Kyoto Japan
Chapter 2

Néstor J. Oviedo
Center for Regenerative
 and Developmental Biology
Forsyth Institute
and
Department of Developmental Biology
Harvard School of Dental Medicine
Boston, Massachusetts, USA
Chapter 12

Ronald J. Tallarida
Department of Pharmacology
Temple University School of Medicine
Philadelphia, Pennsylvania, USA
Email: ronald.tallarida@temple.edu
Chapter 10

Giorgio Venturini
Dipartimento di Biologia
Università di Roma Tre
Viale Marconi
Roma, Italy
Email: venturin@uniroma3.it
Chapters 4, 5, 6

Hiroshi Yamamoto
Department of Biophysics
Graduate School of Science
Kyoto University
Kitashirakawa-Oiwake
Sakyo-ku, Kyoto, Japan
Chapter 2

PREFACE

The study of drug action has benefitted greatly from the development and use of in vivo model systems. In model systems, manipulations and observations can be more rigorously controlled and screens of novel therapeutic agents can be more safely conducted.

No single model system provides all of the possible advantages. At one end, mammalian models allow the study of complex behavioral patterns and the most complex of cognitive functioning. At the other end, models using simple organisms such as *C. elegans* allow the application of the most sophisticated and recent molecular biology and other innovative techniques. The major purpose of the present book is to highlight another model—one that we believe occupies a uniquely important position.

Planarians are the lowest form of animal that possess an integrated neural network including cephalic ganglia organized in a manner that many consider to repesent the earliest brain. They also contain many of the identical biochemical substances utilized by mammals as neurotransmitters and 2nd messengers—and studies suggest that planarians utilize these substances in a similar manner. Most importantly, planarians offer: (1) a relatively sophisticated repertoire of behaviors, including learning and memory, and (2) a remarkable capacity of regeneration.

Drug abuse is a multifaceted phenomenon, involving many biochemical, behavioral, and even toxicological aspects (e.g., deleterious effect on prenatal brain development). Adding to the complexity, many drug abusers use multiple drugs (poly-drug abuse). Great strides have been made in this area, but progress toward an understanding of the biochemical bases of drug abuse and of potential treatment is still needed. Human and non-human primate models are of course of tremendous value, but methodological and other complexities and issues preclude their exclusive use. Insight into certain aspects of the problem can be better, or at least more conveniently, achieved using carefully chosen simpler, yet relevant, alternatives. Planarians represent such an alternative. As shown in this book, planarians exhibit a characteristic and easily quantifiable dose-related withdrawal behavior following abstinence from abused drugs, such as cocaine, and that the effects of dopaminergic, opioid, and other ligands on planarian behavior mimic the characteristics of mammalian pharmacology. In addition, the interplay between systems, as in the practice of poly-drug abuse, can be more easily quantified in this model. In sum, planarians provide a model that can permit insight using a simple yet neurochemically-relevant system that can complement mammalian studies leading to possible clinical therapeutic interventions.

The material presented here provides, we believe, background about the historical development of, and insight into, the 'state-of-the-art' of research into drug action and drug abuse using planarians. It also provides, we believe, insight into future possibilities for creative research into questions of other issues, such as learning and memory (and other complex behaviors) and regeneration (and the discovery of potential pharmacotherapeutic agents). As such, the book should be of interest to students and the general reader as well as specialists. Toward this end, we have recruited authors who are intimately involved in the development and use of planarian model systems. Each chapter has been written in such a way that it can be read independently from the others, but with a uniformity that allows smooth transition from one chapter to the next. It is our desire that this book provides the reader with an opportunity to quickly learn about the use of planarians and some of the results obtained so far. It is also our desire that the book provides an impetus for the 're-discovery' of planarians as a vibrant and valuable research tool.

It seems somehow appropriate that the editors are at Temple University in Philadelphia, because to our knowledge the first study of drug abuse phenomena in planarians was conducted at Temple University by Herbert Needleman during the 1960s.

We would like to thank Timothy Shickley, PhD for suggesting Planaria as a model system and acknowledge the support of NIH grants R01-DA15378 (to RBR) and DA-022694 (to SMR).

Robert B. Raffa, PhD
Scott M. Rawls, PhD

CHAPTER 1

Planaria:
Short Introduction
Robert B. Raffa*

Abstract

Planarians[a] are free-living, nonparasitic, bilaterally symmetrical flatworms (dorsoventrally flattened). They range in size from about 3 to about 15 mm and are found throughout the world. There are several features that make planarians particularly noteworthy and valuable for pharmacologic study, including being the earliest extant example of cephalization (centralized 'nervous system'), capacity for rather complex behavior ('learning and memory') and ability to regenerate.

General Features

There are thousands of water- and land-dwelling planarians; an example of *Dugesia doroto-cephala* is shown in Figure 1.

Planarians have a solid body, but no body cavity (i.e., they are 'acoelomate'). They live in water or on land under rocks and other cover, where they feed upon smaller invertebrates ('zoophagus'), detritus ('detrivore') or decaying organic matter, or diatoms within their ecosystems. Planarians have 'auricles', sensory neurons that project from the sides of their head and contain chemoreceptors that help in the location of food. Their digestive system consists of a mouth (located in the center of the underside of the body), a pharynx and a gastrovascular cavity that branches throughout the body supplying nutrients from food sources to reach all portions of the body. Digestive enzymes secreted from the mouth begin the digestive process. The process continues as the food passes through the pharynx into the gastrovascular cavity for distribution throughout the body.

The planarian excretory system consists of multiple tubes with 'flame' cells and excretory pores distributed along them. The flame cells direct unwanted material to ducts that lead to the excretory pores and subsequent release on the planarian's dorsal surface. Oxygen uptake and carbon dioxide elimination occur by diffusion (possible due to the high surface to volume ratio).

Two eyespots ('ocelli') are located at the anterior portion of many planarians and function as photoreceptors. The eyespots detect light and intensity gradient and are used to direct the planarian toward the dark. The action spectrum and the spectral sensitivity of the planarian ocellus is maximal at approximately 508 nm, suggesting that the visual pigment in the planarian ocellus is a rhodopsin-like compound.[1] Under each eyespot is a ganglion (cephalization). Two nerve cords ('longitudinal nerves') originate from each ganglion, traverse down the body and connect at the

---

[a]Planaria has become a common name for the genus *Planaria* and sometimes for the individual members of the genus. The term 'planarians' is used here for the latter purpose.

*Robert B. Raffa—Department of Pharmaceutical Sciences, Temple University School of Pharmacy, 3307 N. Broad Street, Philadelphia, Pennsylvania, USA.
Email: robert.raffa@temple.edu

Planaria: A Model for Drug Action and Abuse, edited by Robert B. Raffa and Scott M. Rawls.
©2008 Landes Bioscience.

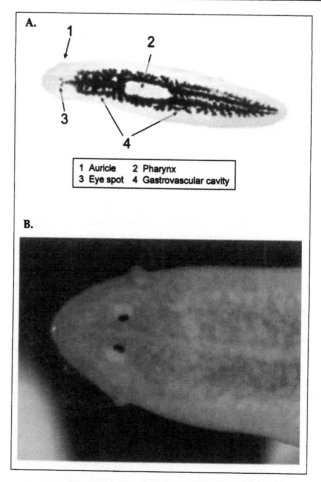

Figure 1. The planarian *Dugesia dorotocephala*. Photomicrographs of (A) slide obtained from Carolina Biological Supply (Burlington, NC) and (B) supplied by Zhe Ding in the author's laboratory.

tail, yielding a primitive bilateral 'spinal cord'. 'Transverse nerves' connect the nerve cords, giving a ladder-like appearance.

Planarians locomote by means of cilia located on the ventral dermis, allowing it to glide along on a film of mucus, or by means of undulations of the whole body initiated by coordinated contractions of muscles built into the body wall.

Reproduction

Planarian strains can reproduce exclusively asexually ('fission'), exclusively sexually, or by both mechanisms.[2-4] Those that reproduce exclusively by only one means are genetically different from strains of the same species capable of reproducing both sexually and asexually. For example,[5] planarians that belong to different asexual strains of *Girardia* (formerly *Dugesia*) *tigrina* and of *Dugesia japonica* are triploid or mixoploid, whereas sexual strains are diploid.[3,4]

Asexual reproduction can occur under the proper conditions of food supply and ambient temperature (above about 10°C, maximum at about 25-28°C).[6] The planarians's body pinches at about mid-length and continues until complete separation is achieved. Each half then forms into an independent whole worm.

Unlike planarians of asexual strains, mature planarians of sexual strains have, sometimes seasonally-dependently, well-defined internal (testes and ovaries) and external reproductive organs (i.e., they are 'hermaphroditic'). During sexual-type reproduction, both planarian partners exchange sperm. Eggs develop within the body and are shed in oval capsules (fertile cocoons) about 5 mm long. The young emerge (hatch) weeks later as juveniles.

Regeneration

Planarians possess an extraordinary ability to regenerate. This involves the generation of new tissue at the wound site by processes incorporating both cell proliferation ('blastema formation') and remodeling of pre-existing tissues to restore the planarian's symmetry and proportion ('morphallaxis').[7] Neoblasts, a proliferative cell population, are central to the ability of planarians to regenerate.[7] The topic of planarian regeneration is covered in Chapter 12.

Conclusion

Planarians provide a combination of simplicity of manipulation with relative complexity of structure and function—including cephalization, response to drugs and regenerative capacity—that make them an important target of study. To this, an additional level of utility and value will now be added with the sequencing of the genome of *Schmidtea mediterranea*.[8]

References

1. Brown HM, Ito H, Ogden TE. Spectral sensitivity of the planarian ocellus. J Gen Physiol 1968; 51:255-260.
2. Benazzi M, Benazzi-Lentati G. Platyhelminthes. In: John B, ed. Animal Cytogenetics. Berlin: Gebruder Borntraeger, 1976:1-182.
3. Ribas M, Riutort M, Baguna J. Morphological and biochemical variation in populations of Dugesia (G.) tigrina (Turbellaria, Tricladida, Paludicola) from the western Mediterranean: Biogeographical and taxonomical implications. J Zool Lond 1989; 218:609-626.
4. Bessho Y, Tamura S, Hori H et al. Planarian mitochondria sequence heterogeneity: relationships between the type of cytochrome c oxidase subunit I gene sequence, karyotype and genital organ. Molec Ecol 1997; 6:129-136.
5. Shagin DA, Barsova EV, Bogdanova EA et al. Identification and characterization of a new family of C-type lectin-like genes from planaria girardia tigrina. Glycobiol 2002; 12:463-472.
6. http://animaldiversity.ummz.umich.edu/site/accounts/information/Turbellaria.html. Accessed 2008.
7. Reddien PW, Alvarado AS. Fundamentals of planarian regeneration. Ann Rev Cell Develop Biol 2004; 20:725-757.
8. http://genome.wustl.edu/genome.cgi?GENOME=Schmidtea%20mediterranea and SECTION=research. Accessed 2008.

CHAPTER 2

Brain and Neural Networks

Kaneyasu Nishimura, Hiroshi Yamamoto, Yoshihisa Kitamura
and Kiyokazu Agata*

Abstract

Freshwater planarians have a relatively well-organized central nervous system (CNS), which consists of the brain and the ventral nerve cords (VNCs). Recently, several neural marker genes have been isolated by expressed sequence tag (EST) projects and DNA chip analysis and extensive molecular anatomical studies have been conducted using *Dugesia japonica*. Brain patterning and domain structure along the anterior-posterior axis and medial-lateral axis are regulated by *wnt*-family genes and *Otx/otp*-related genes, respectively, as in vertebrate brains. Neural connection patterns of each domain have been extensively analyzed by fluorescence dye tracing and visualization of each axon with monoclonal antibodies. In addition, several neuronal subtypes—such as dopaminergic, serotonergic and GABAergic—have been identified by whole mount in situ hybridization using genes for rate-limiting enzymes as probes and their axonal projections have been visualized.

Introduction

Freshwater planarians have a simple body plan with cephalization, a dorso-ventral axis and bilateral symmetry and are thought to be primitive animals that acquired a central nervous system (CNS) at an early stage of evolution. To understand the structure and function of the planarian CNS, we used extensive molecular biological approaches and histological methods, including whole-mount in situ hybridization and immunostaining to study *Dugesia japonica*. The first neural-specific marker gene, *DjPC2*, was isolated by a small scale EST.[1] Then the first monoclonal antibody specific to visual neurons was obtained by random screening of hybridomas against planarian antigens.[2] The domain structure of the planarian brain was identified using *Otx/otp*-related genes as probes.[3,4] Thereafter, an expressed sequence tag (EST) project and DNA chip screening were performed to systematically isolate neural specific genes.[5,6] These studies revealed the planarian brain to be composed of many functionally and structurally distinct domains and sub-domains.[7] To investigate how each domain is connected with other domains, we raised antibodies to representative neurons and their axons.[2,8-10] To trace axonal projections in more detail, we stained individual neurons using DiI or DiD labeling.[11] Recently, we cloned genes encoding rate-limiting enzymes for neurotransmitter synthesis and succeeded in identifying dopaminergic, serotonergic and GABAergic neurons. We raised monoclonal antibodies against these rate-limiting enzymes to trace each neuron type.[12-14]

Recently, RNA interference (RNAi)-based phenotypical analyses were developed to elucidate neural functions.[15,16] We also developed a unique conditional gene knockdown method named 'regeneration-dependent conditional knockdown' (Ready-knock) to unravel the neural circuits in the planarian CNS.[17] In this chapter, we introduce the basic structure of the planarian brain and neural networks as revealed by these studies.

*Corresponding Author: Kiyokazu Agata—Department of Biophysics, Graduate School of Science, Kyoto University, Kitashirakawa-Oiwake, Sakyo-ku, Kyoto 606-8502, Japan. Email: agata@mdb.biophys.kyoto-u.ac.jp

Planaria: A Model for Drug Action and Abuse, edited by Robert B. Raffa and Scott M. Rawls.
©2008 Landes Bioscience.

Gross Structure of the Planarian CNS

To understand the structure of the planarian CNS, we first identified the *DjPC2* (*D. japonica pro-hormone convertase 2* homologue) gene as a pan-neural marker gene in planarians by a small scale EST project[1] (Fig. 1A). By performing whole-mount in situ hybridization using a *DjPC2* antisense probe, the structure of the planarian CNS was revealed. Thereafter, we raised anti-DjPC2 antibody to reveal the anatomy of neural networks in more detail.[11] The *Djsyt* gene (*D. japonica synaptotagmin* homologue) was also identified and used to reveal the neural connections and axonal projections of the planarian CNS.[8] In addition, an antibody against DjSYT specifically visualized neural networks and axonal projections in the planarian CNS (Fig. 1B).

Whole-mount in situ hybridization and immunostaining of DjPC2 and DjSYT enabled us to visualize the planarian CNS, which was composed of the brain and a pair of longitudinal ventral nerve cords (VNCs) (Fig. 1A,B). The brain is an inverted U-shaped structure consisting of two main lobes with nine lateral branches each. The VNCs are located on the ventral side of the body and connected to each other by transverse bipolar/multipolar commissure neurons. The brain is an independent structure from the VNCs and is located on the dorsal side of the VNCs. The eyes consist of two independent cell types: optic nerves and pigment cells and are located on the dorsal side of the head. The optic nerves form an optic chiasma and project to the dorsal-medial side of the brain (Fig. 1C). The nine lateral branches of the brain project to the lateral surface of the head region and appear to have important functions as a chemosensory system (Fig. 1D).

Domain Structure of the Planarian Brain

The planarian brain is composed of several structurally distinct domains that are distinguished by the expression patterns of various genes. Umesono et al demonstrated that three *Otx/otp*-related genes—*DjotxA*, *DjotxB* and *Djotp*—were evolutionarily conserved and expressed in discrete domains in the planarian brain[3,4] (Fig. 2A-C). The expression patterns of these *Otx/otp*-related genes revealed that the planarian brain can be divided into four structural domains: *DjotxA* is expressed in the optic nerves and medial region of the brain, which form a photosensory domain; *DjotxB* is expressed at the main lobes of the brain, which form a signal-processing domain containing several kinds of interneurons; and *Djotp* is expressed in the lateral branches, which form a chemosensory domain. The lateral side of the head region, where *Otx/otp* expression is not detected, contains mechanosensory neurons. Recently, Kobayashi et al demonstrated that *DjwntA* and *DjfzA* contribute to antero-posterior patterning of the planarian brain[18] (Fig. 2D,E). Although *DjotxA*, *DjotxB* and *Djotp* genes are expressed medio-laterally, *DjwntA* and *DjfzA* genes are expressed antero-posteriorly in the brain (Fig. 2F,G). *Wnt* family genes and *Otx/otp* family genes play important roles in domain formation in the planarian brain.

The expression patterns of neural-specific genes that were obtained via EST projects and DNA chip screening also suggest that the planarian brain is functionally divided into several different domains.[5-7] Interestingly, the expression patterns of the three *Otx/otp*-related homeobox genes basically correspond to these functional domains.

Visualization of Neural Projections Using Fluorescent Dyes

How do neurons in the several domains project and connect to each other? In order to answer this question, Okamoto et al established a fluorescent dye injection method and succeeded in tracing neural projections in several domains in the planarian CNS.[11] The optic nerves form an optic chiasma and project to the dorsal-medial side of the brain. In addition, DiI/DiD fluorescent dye injection analysis showed that the optic nerves can be separately traced and project to both the ipsilateral and contralateral sides of the brain with the optic chiasma (Fig. 3B). In addition, some axons of the optic nerves project to the opposite eye. Collectively, the axons of the optic nerves project in three directions: the contralateral side of the brain, the ipsilateral side of the brain and the opposite eye (Fig. 3C,D). Transverse commissure neurons and lateral axonal projections are strongly stained by injection of DiI into the VNCs (Fig. 3E). Almost all VNC neurons form neural networks independently of the brain (Fig. 3F), but some VNC neurons project to the ventral region of the brain. Although the majority of VNC neurons are connected to each other, some VNC neurons connect directly to the brain.

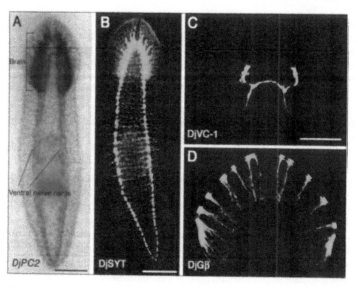

Figure 1. Please see legend on next page.

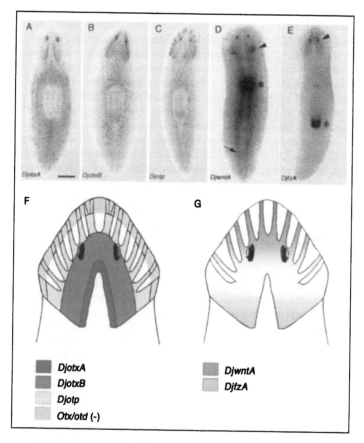

Figure 2. Please see legend on next page.

Figure 1, viewed on previous page. Gross structure of the planarian CNS. A) The distribution pattern of DjPC2 (a planarian pro-hormone convertase 2 homologue)-expressing cells by in situ hybridization. B) Immunostaining with anti-DjSYT (a planarian synaptotagmin homologue) antibody. Transverse axonal projection is clearly visualized. C) Optic nerves stained by anti-DjVC-1 antibody. D) Lateral branches stained by anti-G protein β subunit antibody. Scale bar: 500 μm (A, B), 200 μm (C).

Figure 2, viewed on previous page. Domain structure of the planarian brain. The planarian brain can be divided into at least four structurally and functionally different domains, which are defined by the discrete expression of *Otx/otp* family and *wnt* family genes. A) Both photosensory neurons and their target region in the brain are defined by *DjotxA* expression. B) The two main lobes, where a variety of interneurons have been identified, are defined by *DjotxB* expression. C) Lateral branch neurons are defined by *Djotp* expression and are composed of chemosensory neurons. Mechanosensory neurons are formed in the peripheral region of the head, which is defined by being devoid of the *Djotx* expression of the three *Otx/otp* (–) related genes. D) *DjwntA* was detected in the posterior region of the brain (arrowhead), the VNCs (arrow) and the proximal region of the pharynx (asterisk). E) *DjfzA* showed a complementary expression pattern to *DjwntA*. It was expressed in the anterior region of the brain (arrowhead) and distal region of the pharynx (asterisk). Scale bar: 500 μm (A). F) Schematic drawing of *Otx/otp* family genes. G) Schematic drawing of *wnt* family genes.

Visualization of Neural Networks by Immunofluorescence

We raised several antibodies in order to stain individual neurons in the planarian CNS. These antibodies provide additional tools for use in planarian neuroscience research (Table 1) and are also useful tools for investigation of protein expression levels and observation of the neuro-morphological changes after RNAi experiments.[9,10,17,19,20] To visualize the gross structure of the planarian CNS, several kinds of antibodies, such as anti-DjPC2 antibody, anti-DjSYT antibody and anti-cytochrome b_{561} antibody, were raised as pan-neural markers.[1,8,21] anti-DjVC1 antibody and anti-DjGβ antibody were also raised as a marker for optic nerves and lateral branches of the brain, respectively.[2,10] We recently raised specific antibodies against rate-limiting enzymes for the biosynthesis of neurotransmitters in the planarian CNS and characterized functional neural subtypes such as dopaminergic, serotonergic and GABAergic neurons.[12-14] Because tyrosine hydroxylase (TH) is the rate-limiting enzyme for dopamine biosynthesis, anti-DjTH antibody is a useful tool for labeling of dopaminergic neurons. In the same way, tryptophan hydroxylase

Table 1. Some antibodies for the planarian CNS

Antibody	Homology	Distribution Pattern	Reference
DjPC2	Pro-hormone convertase2	Pan-neural	Agata et al (1998)
DjSYT	Synaptotagmin	Pan-neural	Tazaki et al (1999)
Cytochrome b_{561}	Cytochrome b_{561}	Pan-neural	Asada et al (2002)
DjVC-1	β-arrestin	Optic nerve	Sakai et al (2000)
DjCAM	Cell adhesion molecule	Pan-neural	Fusaoka et al (2006)
DjGβ	G protein β-subunit	Lateral branches	Inoue et al (2007)
DjTH	Tyrosine hydroxylase	DA neurons	Nishimura et al (2007a)
DjAADC-A	Aromatic amino acid decarboxylase	DA neurons and gut	Nishimura et al (2007a)
DjTPH	Tryptophan hydroxylase	Serotonergic neurons	Nishimura et al (2007b)
DjGAD	Glutamic acid decarboxlase	GABAergic neurons	Nishimura et al (2008)

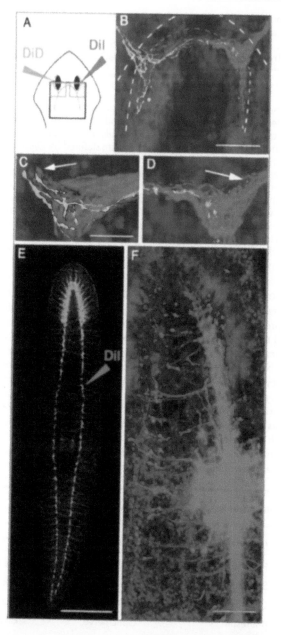

Figure 3. Neural networks visualized fluorescent dyes. A) Schematic drawing of injection sites in the two sides with DiI (megenta) and DiD (green). B) The projection of optic nerves of the two sides labeled by DiI and DiD. Both optic nerves showed ipsilateral and contralateral projections. C) Higher magnification view of left side. D) Higher magnifiecation view of right side. These panels show that some of the contralateral projections are also extend directly to the opposite eye (arrow). E) Image of DiI injection in the trunk region. F) Image obtained by the head to the trunk region obtained by DiI injection. Nerves form bundles from the VNC to the head and some cell bodies are located in the VNC commissures and head region (Okamoto et al, 2005). Scale bar: 100 μm (B, F), 50 μm (C), 500 μm (E). Reproduced from: Okamoto K et al. Zoolog Sci 2005; 22:535-546;[11] with permission from the Zoological Society of Japan.

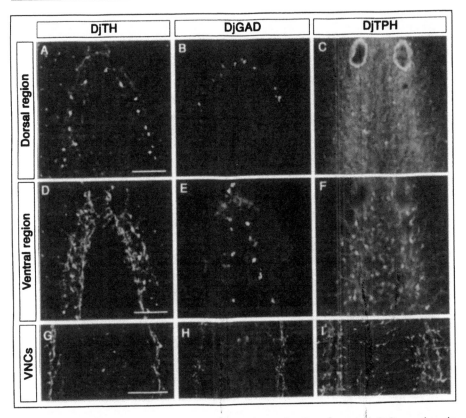

Figure 4. Immunofluorescence of several functional neural cell markers. A,D,G) Dopaminergic neurons stained by anti-DjTH antibody. B,E,H) GABAergic neurons stained by anti-DjGAD antibody. C,F,I) Serotonergic neurons stained by anti-DjTPH antibody. (A-C) Dorsal and (D-F) ventral view of the head. (G-I) Ventral nerve cords in the trunk region. Scale bar: 100 μm (A,D,G).

(TPH) and glutamic acid decarboxylase (GAD) are the rate-limiting enzymes for serotonin and GABA (γ-aminobutyric acid) biosynthesis, respectively. Anti-DjTPH antibody and anti-DjGAD antibody label serotonergic neurons and GABAergic neurons, respectively (Fig. 4).

Dopaminergic neurons can be classified into several types according to their distribution in the brain. They are located in the dorsal and ventral sides of the brain and in the lateral sides of the head (Fig. 4A,D) and are aligned in an inverted U-shape and connected with each other in the dorsal side of the brain (Fig. 4A). GABAergic neurons can be classified into two types based on distribution (Fig. 4B,E). They are located in the dorsal and ventral sides of the brain. GABAergic neurons in the dorsal side are aligned in an inverted U-shape and connected with each other (Fig. 4B), whereas GABAergic neurons in the ventral side are connected with each other by transverse axonal projections (Fig. 4E). In the trunk region, the axonal projection of DjGAD-immunopositive neurons was detected, but not cell bodies in the VNCs (Fig. 4H). Serotonergic neurons are widely distributed along the VNCs. They form a multipolar structure and are connected with each other by transverse axonal projections at the VNCs (Fig. 4I). DjTPH-immunoreactivity was detected at eye pigment cells (Fig. 4C).

As revealed by the immunostaining with these antibodies, each neuron had a specific neural cell morphology and was part of a specific neural network in the planarian CNS. We succeeded in visualizing these neural networks at the cellular level. Previously, Asami et al performed small-scale single-cell level gene profiling in the planarian brain using by fluorescence activated cell sorting (FACS) and semi-quantitative PCR analysis.[22] These studies have been summarized by constructing a 3D brain model at the single cell level, which is available on the web (http://www.brh.co.jp).

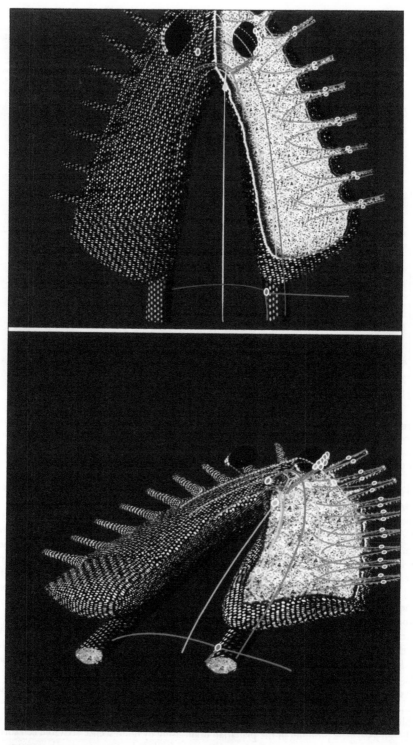

Figure 5. Please see legend on next page.

Figure 5, viewed on previous page. Summary of neural networks in the planarian CNS. Optic nerves project in three directions: (1) ipsilateral side (green); (2) contralateral side: the medial region (orange); and (3) contralateral side: to the opposite eye (orange). Lateral branch neurons project to their stump region (pink). Commissural neurons connect the left and right areas through dorsal arc regions (blue). Pharynx nerves connect with dorsal arc region 2 (light green). Brain and VNC have multipolar neurons (purple). Reproduction from: Okamoto K et al. Zoolog Sci 2005; 22:535-546;[11] with permission from Zoological Society of Japan.

Conclusions

Although the planarian brain is often described as a simple structure, it includes a large number of neural cell types, domain structures and neural circuits. These neural networks and connections revealed by immunostaining and fluorescent dye injection are summarized in Figure 5. Knowledge about morphology, localization and organization is an essential first step for understanding neural functions. In the last decade, histological analyses, including whole-mount in situ hybridization and immunohistochemistry, have been established and several bioinformatics techniques, including EST projects and genome sequencing, have been applied to planarian research.[3,5,11,23,24] In addition, several functional analysis methods such as conditional RNA interference (Ready-knock method), pharmacological approaches and behavioral assay systems, have been established.[12,17,20] Thus, various tools and techniques are available for examining the morphology and function of the planarian CNS at the molecular and cellular levels. Planarians have begun to receive much attention in the field of stem cell research and in neuroscience research in general.[25,26]

References
1. Agata K, Soejima Y, Kato K et al. Structure of the planarian central nervous system (CNS) revealed by neuronal cell markers. Zoolog Sci 1998; 15:433-440.
2. Sakai F, Agata K, Orii H et al. Organization and regeneration ability of spontaneous supernumerary eyes in planarians—eye regeneration field and pathway selection by optic nerves. Zoolog Sci 2000; 17:375-381.
3. Umesono Y, Watanabe K, Agata K. A planarian orthopedia homolog is specifically expressed in the branch region of both the mature and regenerating brain. Dev Growth Differ 1997; 39:723-727.
4. Umesono Y, Watanabe K, Agata K. Distinct structural domains in the planarian brain defined by the expression of evolutionarily conserved homeobox genes. Dev Genes Evol 1999; 209:31-39.
5. Mineta K, Nakazawa M, Cebrià F et al. Origin and evolutionary process of the CNS elucidated by comparative genomics analysis of planarian ESTs. Proc Natl Acad Sci USA 2007; 100:7666-7671.
6. Nakazawa M, Cebrià F, Mineta K et al. Search for the evolutionary origin of a brain: planarian brain characterized by microarray. Mol Biol Evol 2003; 20:784-791.
7. Cebrià F, Kudome T, Nakazawa M et al. The expression of neural-specific genes reveals the structural and molecular complexity of the planarian central nervous system. Mech Dev 2002; 116:199-204.
8. Tazaki A, Gaudieri S, Ikeo K et al. Neural network in planarian revealed by an antibody against planarian synaptotagmin homologue. Biochem Biophys Res Commun 1999; 260:426-432.
9. Fusaoka E, Inoue T, Mineta K et al. Structure and function of primitive immunoglobulin superfamily neural cell adhesion molecules: a lesson from studies on planarian. Genes Cells 2006; 11:541-555.
10. Inoue T, Hayashi T, Takechi K et al. Clathrin-mediated endocytic signals are required for the regeneration of, as well as homeostasis in, the planarian CNS. Development 2007; 134:1679-1689.
11. Okamoto K, Takeuchi K, Agata K. Neural projections in planarian brain revealed by fluorescent dye tracing. Zoolog Sci 2005; 22:535-546.
12. Nishimura K, Kitamura Y, Inoue T et al. Reconstruction of dopaminergic neural network and locomotion function in planarian regenerates. Dev Neurobiol 2007a; 67:1059-1078.
13. Nishimura K, Kitamura Y, Inoue T et al. Identification and distribution of tryptophan hydroxylase (TPH)-positive neurons in the planarian Dugesia japonica. Neurosci Res 2007b; 59:101-106.
14. Nishimura K, Kitamura Y, Umesono Y et al. Identification of glutamic acid decarboxylase gene and distribution of GABAergic nervous system in the planarian Dugesia japonica. Neuroscience 2008; 153:1103-1114.
15. Sánchez Alvarado A, Newmark PA. Double-stranded RNA specifically disrupts gene expression during planarian regeneration. Proc Natl Acad Sci USA 1999; 96:5049-5054.
16. Newmark PA, Reddien PW, Cebrià F et al. Ingestion of bacterially expressed double-stranded RNA inhibits gene expression in planarians. Proc Natl Acad Sci USA 2003; 100:11861-11865.

17. Takano T, Pulvers JN, Inoue T et al. Regeneration-dependent conditional gene knockdown (ready-knock) in planarian: demonstration of requirement for Djsnap-25 expression in the brain for negative phototactic behavior. Dev Growth Differ 2007; 49:383-394.

18. Kobayashi C, Saito Y, Ogawa K et al. Wnt signaling is required for antero-posterior patterning of the planarian brain. Dev Biol 2007; 306:714-724.

19. Cebrià F, Newmark PA. Planarian homologs of netrin and netrin receptor are required for proper regeneration of the central nervous system and the maintenance of nervous system architecture. Development 2005; 132:3691-3703.

20. Inoue T, Kumamoto H, Okamoto K et al. Morphological and functional recovery of the planarian photosensing system during head regeneration. Zoolog Sci 2004; 21:275-283.

21. Asada A, Kusakawa T, Orii H et al. Planarian cytochrome b(561): conservation of a six transmembrane structure and localization along the central and peripheral nervous system. J Biochem 2002; 131:175-182.

22. Asami M, Nakatsuka T, Hayashi T et al. Cultivation and characterization of planarian neuronal cells isolated by Fluorescence Activated Cell Sorting (FACS). Zoolog Sci 2002; 19:1257-1265.

23. Zayas RM, Hernandez A, Habermann B et al. The planarian schmidtea mediterranea as a model for epigenetic germ cell specification: analysis of ESTs from the hermaphroditic strain. Proc Natl Acad Sci USA 2005; 102:18491-18496.

24. Robb SM, Ross E, Sánchez Alvarado A. SmedGD: the schmidtea mediterranea genome database. Nucleic Acids Res 2008; 36:D599-606.

25. Cebrià F. Regenerating the central nervous system: how easy for planarians. Dev Genes Evol 2007; 217:733-748.

26. Agata K, Umesono Y. Brain regeneration from pluripotent stem cells in planarian. Philos Trans R Soc Lond B Biol Sci 2008; 363:2071-2078..

Gene Expression in the Brain and Central Nervous System in Planarians

Katsuhiko Mineta, Kazuho Ikeo and Takashi Gojobori*

Abstract

In planarians, the CNS forms an inverted U-shape consisting of a cephalic ganglion and several branches. Recent advances in molecular studies using expressed sequence tags (EST) and cDNA microarray techniques allow more detailed studies of the planarian central nervous system (CNS). Planarians express a wide range of CNS-related genes, including those essential for functioning of the brain, showing that there are possibly more than seven functional regions in the cephalic ganglion. Although planarians have an apparently primitive and simple morphology, they have a well-organized and complex CNS structure.

Introduction

Due to its highly organized functions and developmental patterning system, the CNS is a key organ for understanding the evolutionary divergence of animals. Planarians possess a CNS with a simple, primitive morphology.[1,2] Although the phylogenetic position of planarians, which belong to the phylum Platyhelminthes, has not yet been completely established, Platyhelminthes is positioned at or near the origin of bilateral animals, based on their morphology and developmental biology[3,4] as well as on recent molecular evolutionary studies.[5] Thus, there is no doubt that planarians are among the descendants of early bilateral animals. For these reasons, the planarian CNS can provide clues about the evolutionary history of the CNS, in particular, that of the brain. In this chapter, we discuss the planarian CNS, based on recent progress using EST sequencing and cDNA microarray analysis.

CNS of Planarians

Planarians possess a CNS with a simple, primitive morphology.[1,2] In brief, the planarian CNS is composed of a cephalic ganglion and a pair of ventral nerve cords. The cephalic ganglion forms an inverted U-shaped structure with nine branches connecting to the sensory organs on either side.[6,7] The planarian cephalic ganglion exhibits many morphological features similar to the vertebrate CNS, such as multi-polarized neurons.[8] Moreover, genes homologous to *otx* are expressed specifically in the planarian cephalic ganglion.[9] Although further experiments and more genetic information are required, we consider that these facts support the classification of the planarian cephalic ganglion as a primitive brain[10,11] and a potential ancestor of the vertebrate brain.[12]

*Corresponding Author: Takashi Gojobori—Center for Information Biology and DNA Databank of Japan, National Institute of Genetics, 1111 Yata, Mishima, Shizuoka 411-8540, Japan. Email: tgojobor@genes.nig.ac.jp

Planaria: A Model for Drug Action and Abuse, edited by Robert B. Raffa and Scott M. Rawls ©2008 Landes Bioscience.

Planarian Homologs of Known Nervous System-Related Genes

Planarian ESTs and the CNS

Neuronal marker genes provide a powerful approach for studying the planarian CNS. The marker genes are the homologs of CNS-related or nervous system (NS)-related genes known to be functional in other organisms, such as humans. EST sequencing is a convenient method for identifying these genes, compared with genome sequencing. In planarians, EST studies have been performed in *Dugesia japonica*,[11] *Schmidtea mediterranea*[13] and other species.[14]

For example, we characterized 3,101 nonredundant EST clones from the planarian head region in collaboration with the group of Professor Kiyokazu Agata at Kyoto University. Since the planarian CNS is located in the head region, these ESTs were expected to have a bias toward CNS-related genes. Among the nonredundant clones, 1,385 clones were homologous to genes with known functions in other organisms.

Diversity of the NS-Related Genes in Planarians

To identify the genes related to CNS function in planarians, Neural System (NS)-related genes were searched. Among the clones corresponding to functionally known genes, we identified 116 clones that exhibited significant similarity (less than 1.0e-4 in E-value) to NS-related genes previously characterized in other organisms. The details were published by Mineta et al.[11] In summary, these NS-related clones could be separated into the following five functional groups: (A) 42 clones were related to neurotransmission, including genes related to neurotransmitters, receptors/channels, synaptic vesicles and their transport; (B) 33 clones were related to the neural network, including the Ig-CAM family, the cadherin family and axon guidance; (C) 21 clones showed homology with genes for brain morphogenesis/neural differentiation such as the BMP cascade, the Wnt cascade, the FGF cascade and the notch cascade; (D) 11 clones were related to sensory systems, including the photosensory, chemosensory and mechanosensory systems; and (E) 9 clones were related to others proteins, such as brain protein AB239. The planarian NS-related genes identified fell into various functional groups known to be important for CNS function. In particular, the clones in category (C) of brain morphogenesis/neural differentiation are known to be essential for CNS development in higher organisms. These NS-related clones in planarians may not all be orthologous to functionally known NS-related genes in other organisms, some of them may instead be paralogous genes. In either case, the planarian sequences were significantly similar to those of functionally known genes in other organisms. Thus, planarians have a wide variety of NS-related genes, indicating the complex molecular composition of the planarian CNS.

Conservation of Planarian NS-Related Genes Among Bilateral Animals and Species without a Nervous System

To study the evolutionary emergence of the gene set and the features of the CNS in various species, comparative genomics was employed in an extensive way. Sequences homologous to the 116 NS-related planarian clones were extensively searched against all ORFs in the complete genome sequences of humans, *Drosophila melanogaster* and *Caenorhabditis elegans*, all of which possess a CNS (Fig. 1). As a result, 110 of the 116 planarian head genes (i.e., more than 95%) were shared by all of the bilateral animals examined (*C. elegans*, *D. melanogaster* and humans) (Fig. 1). These clones included genes from all of the five functional categories noted above. Interestingly, all of the 116 genes examined were completely shared between humans and planarians (Fig. 1), whereas six of these clones were absent in *C. elegans* and/or *D. melanogaster*. The species examined belong to the three different groups of bilateral animals (Deuterostomia, Ecdysozoa and Lophotrochozoa)[5] all of which possess a CNS. The NS-related genes identified appeared to include (i) genes related specifically to the CNS and (ii) other genes related to the NS. Thus, these genes can be considered to be genes essential for the CNS.

These NS-related genes were also compared with all ORFs from the complete genomes of *Saccharomyces cerevisiae* and *Arabidopsis Thaliana,* which do not possess a nervous system. As a result, 30% of the 116 genes were found to be shared between *S. cerevisiae* and planarians and 37%

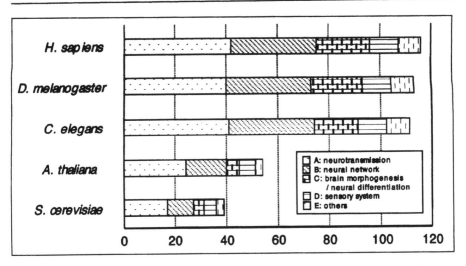

Figure 1. Comparisons between NS-related genes of the planarian and all ORFs in the complete genomes of S. cerevisiae, A. thaliana, C. elegans, D. melanogaster and H. sapiens. The horizontal axis represents the number of sequences homologous to the planarian NS-related clones (116 clones in total) in the genome examined. The functional categories are indicated by different colors. Modified from reference 11.

were shared between *A. thaliana* and planarians (Fig. 1). The functional category with the largest number of shared clones was (A), neurotransmission, which included 17 clones homologous to ORFs from *S. cerevisiae* and 24 clones homologous to ORFs from *A. thaliana*. The time of divergence of *A. thaliana*, *S. cerevisiae* and metazoans has been estimated at about 1,700 million years ago.[15] The emergence of the CNS is generally assumed to be concomitant with the emergence of bilateral animals (before the Cambrian explosion; more than 560 million years ago[16]). Since *A. thaliana* and *S. cerevisiae* do not possess a CNS or nervous system, this finding implies that the evolutionary origin of NS-related genes greatly predates the emergence of the nervous system and CNS. Based on these observations, we speculate that, during evolution, many genes which are now functional in the nervous system or CNS may have been recruited from genes used in unicellular systems. To summarize these results, a possible scenario of the evolution of the CNS based on the NS-related genes is shown in Figure 2.

Planarian cDNA Microarrays and Regionalization of the CNS

Planarian cDNA Microarrays and the CNS

As described above, EST sequences have provided information on the planarian CNS. The comparative study of ESTs can provide insights into CNS function, based on known CNS genes in other organisms. However, there might be other unknown functional genes in the planarian CNS. To investigate this aspect, cDNA microarray analysis can be used to examine the novel genes expressed in the CNS. Nakazawa et al[10] constructed cDNA microarrays carrying 1,640 nonredundant planarian genes. They then conducted large-scale screening of head region-specific genes in the planarian. Competitive hybridization between cDNAs from the head region and the other body region of planarians revealed 205 genes with head region-specific spikes, including genes essential to the vertebrate nervous system.

The 205 genes can be separated into four categories based on a homology search. Eighty-two genes are homologous to functionally known genes. In particular, 33 of the 82 genes are homologous to neural-related genes in other organisms. The remainder of the 123 genes have unknown functions. They therefore found four times as many unknown head region-specific genes as

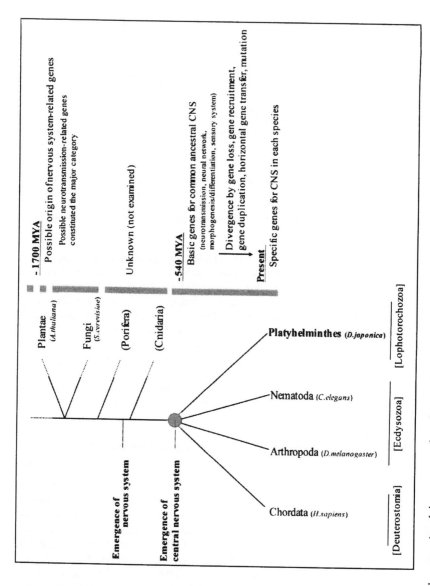

Figure 2. A possible scenario of the process of evolution of the CNS. Possible events during evolution of CNS from the NS-related genes are noted. Modified from reference 11.

known genes through cDNA microarray analysis. The planarian brain contains a large number of functionally unknown genes that are currently beyond our fundamental understanding of the basic neural genes.

Functional Regionalization in the Planarian Brain

Whole mount in situ hybridization for head region-specific genes demonstrated the molecular features of the planarian brain. We focused on the top 30 genes which were highly expressed in the head region, out of the total 205 candidates and observed distinct expression patterns that could be separated into seven categories, designated types A to G (Fig. 3).

Type A genes are expressed throughout the CNS. These genes are likely to have basic and fundamental functions in the nervous system of most organisms. Most of the type A genes show sequence similarities to neural-specific genes in other organisms. Compared with type A genes,

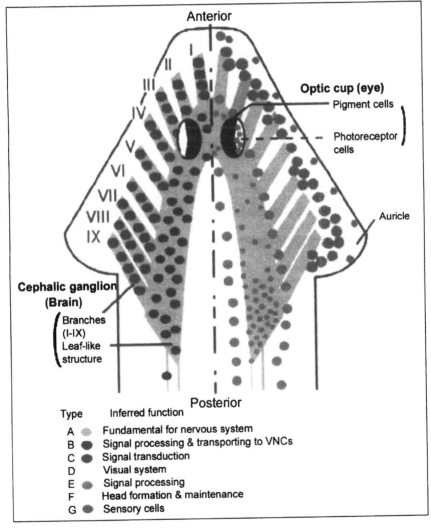

Figure 3. Cytoarchitecture map of planarian brain. The represented patterns of each type (A to G) are schematically drawn using different colors at a half side. Reprinted with permission from reference 10.

type B and C genes are expressed in specific regions of the cephalic ganglion. Type B genes are expressed particularly in the leaf-like regions on both sides of the bilobed cephalic ganglion. As this region is connected to Vertical Nervous Code (VNC), type B genes may have a role in processing and transporting biological signals to the body. Contrary to type B genes, type C gene expressions are restricted to the branch regions in the cephalic ganglion. These genes are likely to be involved in signal transduction and may code proteins such as homolog of the G-protein alpha subunit. These differing expression patterns imply that functional regionalization occurs within the planarian CNS. Since the branch and leaf-like regions are morphologically distinct, these structurally different regions could be expected to possess specialized functions. The expression patterns of the type E genes, however, demonstrate that regionalization occurs even within the leaf-like structure, which appears to be morphologically homogenous. Type E expression patterns indicate regionalization in the planarian cephalic ganglion, similar to that observed in the human brain. The genes categorized in type F show particular patterns in which expression signals appear not only in the cephalic ganglion, but also in the entire head region. Based on the results of RNA interference experiments on these genes that are expressed in the anterior side of the planarian body, we assume that they are strongly related to cephalization in planarians.[17] The last group, type G genes, are expressed specifically in sensory cells. The cells that stain positively for these genes are located in the region extending to the tips of the brain branches, suggesting that these cells are related to the CNS.

The variety of expression patterns of the top 30 head region-specific genes demonstrates the highly organized nature of the planarian CNS. Their expressions indicate that the planarian CNS is functionally well-regionalized. This functional regionalization of the planarian brain allows the performance of complex processes, as observed in higher organisms.

Conclusion

Here, based on gene expression information gained using techniques such as ESTs and cDNA microarrays, we have demonstrated that the planarian CNS is not simple but quite complex. From these approaches based on large-scale molecular analysis, the planarian CNS has been shown to express a wide-variety of NS-related genes, suggesting that they possess essential, CNS-related genes and that the planarian CNS is functionally well-developed. The recent progress of the planarian genome project[18] also gives us an opportunity to examine the planarian CNS from the viewpoint of the genes or gene sets. Moreover, we suggest that methods based on evolutionary studies can play an important role in the study of the CNS. As we noted above, planarians are one of the early descendants of bilateral animals and thus their CNS and brain are considered to reflect ancient features. The evolutionary study of the planarian brain provides insights into the understanding of the complex nervous systems of higher organisms, such as humans. This provides impetus for studies to gain further insights into planarian CNS structure and function.

Acknowledgments

We thank Professor Kiyokazu Agata at Kyoto University for collaboration of this work with us for many years.

References

1. Baguna J. From invertebrates to humans. In: Ferretti P, Geradieu J, eds. Cellular and Molecular Basis of Regeneration. Chichester: John Wiley and Sons Ltd, 1998:135-165.
2. Riger RM, Tyler S, Smith III JPS et al. In: Harrison FW, ed. Microscopic Anatomy of Invertebrates. New York: Wiley-Liss, 1991:7-140.
3. Neilsen C. Animal Evolution, 2nd ed. Oxford: Oxford University Press, 2001:268-280.
4. Hyman LH. In: The Invertebrates II. Platyhelminthes and Rhynchocoela. The acoelomate bilateria. New York: McGraw-Hill, 1951:
5. Aguinaldo AMA, Turbeville JM, Linford LS et al. Evidence for a clade of nematodes, arthropods and other moulting animals. Nature 1997; 387:489-493.
6. Agata K, Soejima K, Kato K et al. Structure of the planarian central nervous system (CNS) revealed by neuronal cell markers. Zoolog Sci 1998; 15:433-440.

7. Okamoto K, Takeuchi K, Agata K. Neural projections in planarian brain revealed by fluorescent dye tracing. Zoolog Sci 2005; 22:535-546.
8. Sarnat HB, Netsky MG. The brain of the planarian as the ancestor of the human brain. Can J Neurol Sci 1985; 12:296-302.
9. Umesono Y, Watanabe K, Agata K. Distinct structural domains in the planarian brain defined by the expression of evolutionarily conserved homeobox genes. Dev Genes Evol 1999; 209:18-30.
10. Nakazawa M, Cebrià F, Mineta K et al. Search for the evolutionary origin of a brain: Planarian brain characterized by microarray. Mol Biol Evol 2003; 20:784-791.
11. Mineta K, Nakazawa M, Cebria F et al. Origin and evolutionary process of the CNS elucidated by comparative genomics analysis of planarian ESTs. Proc Natl Acad Sci USA 2003; 100:7666-7671.
12. Sarnat HB, Netsky MG. When does a ganglion become a brain? Evolutionary origin of the central nervous system. Semin Pediatr Neurol 2002; 9:240-253.
13. Alvarado AS, Newmark PA, Robb SMC et al. The Schmidtea mediterranea database as a molecular resource for studying platyhelminthes, stem cells and regeneration. Development 2002; 129:5659-5665.
14. Ishizuka H, Maezawa T, Kawauchi J et al. The Dugesia ryukyuensis database as a molecular resource for studying switching of the reproductive system. Zoolog Sci 2007; 24:31-37.
15. Nei M, Xu P, Glazko G. Estimation of divergence times from multiprotein sequences for a few mammalian species and several distantly related organisms Proc Natl Acad Sci USA 2001; 98:2497-2502.
16. Futuyma DJ. Evolutionary Biology, 3rd ed. Massachusetts: Sinauer Associates Inc, 1997:165-200.
17. Cebrià F, Kobayashi C, Umesono Y et al. FGFR-related gene nou-darake restricts brain tissues to the head region of planarians. Nature 2002; 419:620-624.
18. Robb SM, Ross E, Alvarado AS. SmedGD: The Schmidtea mediterranea genome database. Nucleic Acids Res 2008; 36(Database issue):D599-606.

CHAPTER 4

Catecholamines in Planaria

Antonio Carolei, Irene Ciancarelli and Giorgio Venturini*

Abstract

Planaria are free-living flatworms and represent the most primitive example of centralization and cephalization of the nervous system along phylogeny. Several neurotransmitters such as dopamine, norepinephrine and serotonin which are present in the human brain have also been identified and investigated in the planaria nervous system. *Dugesia gonocephala* s.l. displays high concentrations of dopamine and lower amounts of norepinephrine. Moreover, the exposure to drugs acting on neural transmission causes specific and stereotyped behavioural patterns. Dopaminergic hyperstimulation rises cAMP and induces typical hyperkinesias, whereas dopaminergic blocking agents decrease cAMP and motility, suggesting similar neurochemical functional mechanisms in invertebrates as in mammals. The dopaminergic-cholinergic balance that was found in the planaria nervous system has been compared to that observed in the extrapyramidal system of mammals. For the above reported reasons, *Dugesia gonocephala* s.l. has been proposed as a reliable model to investigate neurotransmission and related human diseases.

Introduction

Invertebrates are suitable research models in neurobiology and neuropharmacology because of the simple hierarchical organization of their nervous system. Flatworms emerged nearly 40 years ago as a very simple neurobiological system to investigate complex neuronal functions and interactions. Among invertebrates, planaria are the first living organisms in the phylogenetic scale which exhibit centralization and cephalization of the nervous system. Their nervous system consists of a cerebrum and of two symmetric longitudinal cords with ladder-like interconnections representing the first organized stage of cerebral and spinal neurons in invertebrates. Planaria move by means of active muscle contractions and by movements of their cilia which are located on the ventral side. Coordination of behaviour, movement, metabolism and reproduction is performed by the nervous system possibly through a combination of synaptic and paracrine signals. According to the former observations by Welsh and King,[1] dopamine and acetylcholine were described as the operating neurotransmitters in *Dugesia tigrina*, *Procotyla fluviatilis* and *Phagocata oregonensis* since high amounts of dopamine and acetylcholine were found. An active role of the same neurotransmitters was demonstrated in the nervous system of *Dugesia gonocephala* s.l.[2]

In planaria the exposure to different drugs acting on these neurotransmitters causes specific and stereotyped behavioural patterns. The stimulation of the dopamine receptors with L-Dopa or dopaminergic agonists induces hyperkinesias represented by abnormal screw-like movements, whereas dopamine receptors blocking drugs antagonize those behavioural motor patterns. Therefore, since behavioural responses are more easily evaluated in planaria than in other more sophisticated hierarchical animal models, *Dugesia gonocephala* s.l. may provide a simple tool for neuropharmacological investigations aimed to study not only the dopaminergic but also the cholinergic neurotransmission

*Corresponding Author: Giorgio Venturini—Dipartimento di Biologia, Università di Roma Tre Viale Marconi 446, 00146 Roma, Italy. Email: venturin@uniroma3.it

Planaria: A Model for Drug Action and Abuse, edited by Robert B. Raffa and Scott M. Rawls. ©2008 Landes Bioscience.

and to evaluate the pharmacological properties of drugs acting on those systems. In planaria also some peptides such as melanocyte-stimulating hormone release-inhibiting factor (MIF), oxytocin, Met- and Leu-enkephalin exhibit behavioural effects, possibly depending on peptidergic neurotransmission or neuromodulation and/or endocrine-like permissive actions.[3-6]

In the early seventies a basic study was performed to investigate the dopaminergic and cholinergic properties of the planaria nervous system and the effects of some specific drugs acting on those neurotransmitters.[2]

Catecholamines

In the *Dugesia gonocephala* s.l. nervous system are present high concentrations of dopamine and lower amounts of norepinephrine.[7,8] Ultrastructural studies, focusing on the morphological characterization of the synapse, showed the presence of dense-core vescicles, strictly related to the dopamine content.[9] Dopamine modulates the locomotor activity of planaria. The exogenous stimulation of the dopaminergic system causes characteristic abnormal patterns of motility. Modifications in second messenger levels such as an increase in cAMP levels induced by L-dopa and dopamine agonists and the corresponding decrease of the same levels after treatment with dopaminergic blocking agents confirm even in planaria the existing relationship between dopaminergic receptors and adenylate cyclase.[10] The effects of dopamine in planaria are mediated by two types of dopamine receptors, which, on the basis of their pharmacological properties have been identified as D1- and D2-like as compared to dopamine binding sites of mammals. Stimulation of the two dopaminergic receptors causes different locomotor behaviours in the flatworm; D1 stimulation induces screw-like hyperkinesias whereas D2 stimulation triggers a C-like tonic postural attitude. Both responses can be inhibited by selective D1 and D2 blockers.[10] Dopamine agonists increase cAMP levels, an effect that is counteracted by pretreatment with receptor antagonists.[11]

In naïve planaria monoamine contents show some variability, probably due to thermal or seasonal changes, as observed in other invertebrates.[12] On the contrary, behavioural and biochemical responses induced by pharmacological treatments were always constant. Studies with 6-hydroxydopamine (6-OHDA) which decreases dopamine levels confirm that dopamine is fundamental for planaria locomotion. Exposure to 6-OHDA produced significant loss of motility ranging from hypokinesia, bridge-like posture, to complete paralysis.[13] Moreover, benserazide, a specific inhibitor of dopa-decarboxylase, the enzyme which converts L-dopa to dopamine, reduced the motor performances of planaria decreasing the availability of dopamine at the presynaptic level.[14] Apomorphine, a dopamine agonist drug, induced hyperkinesias in planaria, like in mammals, mainly acting on the postsynaptic dopaminergic receptors. The exposure of planaria to haloperidol, a postsynaptic dopaminergic blocking drug, caused a gradual and dose-dependent reduction of movement, up to complete immobility. The increase in motor activity observed after L-dopa exposure is due to an increase in the synthesis and release of dopamine. On the opposite, after reserpine treatment, which impairs neurotransmission, a cataleptic effect is observed as a consequence of catecholamine and indolamine depletion. After L-dopa treatment, together with an increase in dopamine content, a relevant increase in serotonin levels is also observed in planaria.[15] On the other hand, no change of the motor pattern has been observed with clonidine, an α-noradrenergic stimulating agent. 2-Br-α-ergocryptine, used as a dopaminergic agonist in the early treatment of Parkinson's disease, induces a motor pattern similar to the one observed with other dopamine receptor agonists.[2]

Acetylcholine

Acetylcholine (Ach) is the typical neurotransmitter acting at the neuromuscular junction in vertebrates and invertebrates. Among invertebrates, Ach has been implicated as a neuromuscular transmitter primarily excitatory, though it may also have inhibitory effects.[8] In planaria Ach causes hypokinesia leading to prolonged tonic contraction causing an apparent shortening of the entire body length. Exposure to cholinergic antagonists produces moderate hyperkinesias with screw-like movements similar to those induced by dopaminergic agonists. In planaria physostigmine, a cholinesterase-blocking drug, produces hypokinesia with a bridge-like posture.[16] Treatment with atropine, an anticholinergic drug with antimuscarinic properties that competitively antagonizes acetylcholine, induces marked hyperkinesias followed by curling of the animal upon the sagittal plane. This effect might be the consequence of an imbalance of the dopaminergic/cholinergic activity in favour of the former, as in rodents and humans. A similar result was observed after administration of biperiden, an anticholinergic drug.[16]

Serotonin

Serotonin (5HT) is an ubiquitous neuroactive agent. In *Dugesia gonocephala* s.l. 5HT controls planaria regeneration and mediates the regulation of some neuromuscular functions, stimulating motility.[8,11] Binding sites for [³H]lysergic acid diethylamide (LSD), a serotonin receptor agonist, were identified in planaria membranes by ligand binding studies.[17] The 5HT receptor antagonists methiothepin and dihydroergocriptine were also efficient displacers of [³H]LSD binding. When planaria were decapitated in the presence of 5HT antagonists, head regeneration was retarded.[17] These results suggest that planaria possess serotoninergic binding sites which are involved in the process of regeneration.[17] The presence of putative 5HT receptors in planaria has been suggested on the basis of PCR experiments.[17]

Glutamate

The acidic amino acids, in particular glutamate, are major excitatory neurotransmitters in both vertebrates and invertebrates. The fast excitatory effect of glutamate is mediated by direct gating of cation channels. Three major classes of ionotropic receptors were identified according to their agonist specificities as NMDA (N-methyl-D-aspartate), AMPA (α-amino-3-hydroxy-5-meth yl-4-isoxazole-propionic acid) and kainate receptors.[8] In planaria, a method rapidly measuring glutamate and aspartate, utilized a straightforward extraction high-pressure liquid chromatography and fluorescence detection. Recently by using a rapid and efficient method for extraction and measurement of glutamate and aspartate relevant levels of these excitatory amino acids have been demonstrated in planarians.[18] This result suggests that in the flatworm glutamate may play a neurotransmitter role.

Histochemistry Studies

Histochemical studies with fluorescence methods were performed to evaluate the modifications occurring in planaria and after treatment with several drugs.[2]

By using fluorescent methods[19-20] in naïve planaria, at the cephalic level many positive irregular star-like cells were visualized, the greater part of which occur at the periphery of an approximately median structure, located between the eyes and the proximal cephalic intestinal branch.[2] As in other animal species, the fluorescence attributable to catecholamines, increased after administration of L-dopa depending on increased dopamine content or after admistration of benserazide which also contributed to increase the L-dopa content.[2] Fluorescence did not change, possibly depending on time of exposure and on the employed dose, upon treatment with dopaminergic agonists or antagonists and was reduced in reserpine-treated planaria, according to its depleting action of monoamines content. These results confirm that in planaria and mammals catecholamine levels respond to pharmacological treatment in a similar way.

Conclusions

Planaria neurons, even if located at the first level of the phylogenetic scale from the ultrastructural and molecular standpoint are basically comparable with those of vertebrates. The observations of histological sections of planaria and the changes observed in tissue fluorescence according to different treatments confirm that dopaminergic neurotransmission is operative in its nervous system. Moreover, an interaction between the dopaminergic and the cholinergic systems was demonstrated suggesting that acetylcholine may also contribute to prevent dopamine-induced hyperkinesias.[21]

In the mid-seventies our pioneer works found that the motor system of *Dugesia gonocephala* s.l. showed striking similarities with the extrapyramidal system of vertebrates and of humans with the evidence of close correlations between dopaminergic and cholinergic neurons.[24] We predicted that the utilization of this model easily examinable and punctually responsive to several drugs, would have been useful in testing drugs which act on the dopaminergic/cholinergic neurotransmission and in discriminating interferences of other pathways, other levels of action and other neurotransmitters. Up to now, neuropharmacological studies in this very simple animal model confirm that the model is still promising to investigate not only basic neuropharmacological aspects but also to speculate on specific and complex mechanisms underlying different human diseases.

Acknowledgments

The authors are indebted to Professor Guido Palladini for his always excellent support in performing most of their studies.

References

1. Welsh JH, King EC. Catecholamines in planarians. Comp Biochem Physiol 1970; 36:683-688.
2. Carolei A, Margotta V, Palladini G. Proposal of a new model with dopaminergic-cholinergic interactions for neuropharmacological investigations. Neuropsychobiology 1975; 1:355-364.
3. Venturini G, Carolei A, Palladini G et al. Peptide-monoamine interactions in planaria and hydra. In: Stefano GB, ed. CRC Handbook of Comparative Opioid and Related Neuropeptide Mechanisms. Boca Raton: CRC Press, Inc, 1986; 1:245-254.
4. Polleri A, Carolei A, Fazio C. A simple invertebrate model as a tool for critical evaluation of behavioural MIF effects in neuropsychiatry. In Proc 2nd World Congr Biol Psychiat Barcelona 1978:135.
5. Palladini G, Medolago Albani L, Margotta V et al. The pigmentary system of planaria II. Physiology and functional morphology. Cell Tissue Res 1979; 199:203-211.
6. Venturini G, Carolei A, Palladini G et al. Naloxone enhances cAMP levels in planaria. Comp Biochem Physiol 1981; 69C:105.
7. Joffe BI, Kotikova EA. Distribution of catecholamines in turbellarians. In: Sakharov DA, Winlow W ed. Simpler nervous systems. Studies in neuroscience. Manchester: Manchester University Press 1991; 13:77-112.
8. Ribeiro P, El Shehabi F, Patocka N. Classical transmitters and their receptors in flatworms. Parassitology 2005; 131:S19-S40.
9. Palladini G, Margotta V, Carolei A et al. The cerebrum of dugesia gonocephala s.l. Platyhelminthes, turbellaria, tricladida. Morphological and functional observations. J Hirnforsch 1983; 24:165-172.
10. Palladini G, Ruggeri S, Stocchi F et al. A pharmacological study of cocaine activity in planaria. Comparative Biochemistry and Physiology 1996; 115C:41-45.
11. Algeri S, Carolei A, Ferretti P et al. Effects of dopaminergic agents on monoamine levels and motor behaviour in planaria. Comp Biochem Physiol 1983; 74C:27-29.
12. Stefano GB, Catapane EJ. Enkephalins increase dopamine levels in CNS of a marine mollusc. Life Sci 1979; 24:1617-1621.
13. Caronti B, Margotta V, Merante A et al. Treatment with 6-hydroxydopamine in planaria (Dugesia gonocephala s.l.): morphological and behavioural study. Comparative Biochemistry and Physiology Part C 1999; 123:201-207.
14. Burkard WP, Gey KF, Pletscher A. A new inhibitor of decarboxylase of aromatic aminoacids. Experientia 1962; 18:411-412.
15. Algeri S, Carolei A, Ferretti P et al. Effects of dopaminergic agents on monoamine levels and motor behaviour in planaria. Comp Biochem Physiol 1983; 74C:27-29.
16. Buttarelli FR, Pontieri FE, Margotta V et al. Acetylcholine/dopamine interaction in planaria. Comp Biochem Physiol C Toxicol Pharmacol 2000; 125:225-231.
17. Saitoh O, Yuruzume E, Nakata H. Identification of planarian serotonin receptor by ligand binding and PCR studies. Neuroreport 1996; 8:173-178.
18. Rawls SM, Gomez T, Stagliano GW et al. Measurement of glutamate and aspartate in planaria. Journal of Pharmacological and Toxicological Methods 2006; 53:291-295.
19. Falck B, Hillarp NA, Thieme G et al. Fluorescence of catecholamine and related compounds condensed with formaldehyde. J Histochem Cytochem 1962; 10:348-354.
20. Fuxe K, Jonsson G. The histochemical fluorescence method for the demonstration of catecholamine. J Histochem Cytochem 1973; 21:293-311.
21. Venturini G, Stocchi F, Margotta V et al. A pharmacological study of dopaminergic receptors in planaria. Neuropharmacology 1989; 28:1377-1382.

CHAPTER 5

Opioids in Planaria

Antonio Carolei, Marco Colasanti and Giorgio Venturini*

Abstract

Planaria, considered the ancestor of all Bilateria, including Vertebrates, represents the most primitive example of centralization and cephalization of the nervous system. A large body of research performed in the last decades suggest a strong conservation of the basic neurochemical mechanisms during evolution. In particular, the flatworm nervous system employs a wide repertoire of neuroactive molecules, including the classical neurotransmitters and several peptides.

The presence of opioid peptides and of the related binding sites, functionally similar to the κ subtype of mammals is well supported in planaria, both on the basis of immunochemical evidence and of pharmacological approaches. A functional interaction between the opioid and dopaminergic system is operating in planaria, with an enkephalinergic modulation of dopamine release.

Abstinence-induced withdrawal phenomena have been described in planaria and the neurotransmitter system similar to the mammalian one, suggests that this flatworm may be particularly suitable as a model for dependence studies.

Opioids

Evidence for the use of opium by humans is present since 3,400 B.C. in the ancient Mesopotamia. The opium poppy was referred to as the 'joy plant' by Sumerians and its euphoric, narcotic and antidiarrhoic properties were well known in the ancient civilizations.

Despite the long history of opiate use and the great amount of studies on their pharmacological properties, a good comprehension of their mechanisms of action was achieved only since 1973, when the presence in the nervous tissue of a specific opioid (opiate-like) receptor was demonstrated. In fact, Pert and Snyder found that tritiated naloxone, an opiate antagonist, specifically binds to an opioid receptor present in the mammalian brain and in the guinea pig intestine.[1] Competition for this binding by various opioids and their antagonists closely paralleled their pharmacological properties. The authors suggested that opioid receptors are confined to the nervous tissue.

In 1975, an endogenous ligand for the opioid receptor was found in mammalian brain and the physiological role of this neuromodulatory system was finally clarified. Hughes obtained and purified from brain tissue a low molecular weight substance acting as an endogenous mediator at central opioid receptor sites.[2] This substance inhibits neurally evoked contractions of the mouse vas deferens and guinea pig myenteric plexus. The described inhibitory action of this substance was antagonized by opioid antagonists such as naloxone or naltrexone at nanomolar concentrations. The purified substance had no effect on the guinea pig and rabbit vas deferens, tissues which do not possess opioid receptor sites. Further studies demonstrated the presence of opiate-like peptides, such as enkephalins and endorphins, in the nervous tissue of several vertebrates and the existence of several classes of opioid receptors, characterized by differential affinity for different endogenous or synthetic ligands.[3,4]

*Corresponding Author: Giorgio Venturini—Dipartimento di Biologia, Università di Roma Tre, Viale Marconi 446, 00146 Roma, Italy. Email: venturin@uniroma3.it

Planaria: A Model for Drug Action and Abuse, edited by Robert B. Raffa and Scott M. Rawls.
©2008 Landes Bioscience.

The Opioid System in Invertebrates

After the demonstration of endogenous opioids and of their specific receptors, the pioneer studies on the phylogenetic distributions seemed to suggest that the opioid system was present only in vertebrates.[5,6] The authors concluded that during the course of evolution the vertebrate nervous system acquired a qualitatively new type of synaptic function, responsible for opioid receptor interactions.

Starting in 1978, increasing evidence suggested the presence in several invertebrates of both opioid receptors and endogenous opioid ligands, thus further demonstrating how early in the evolution these receptors and ligands have appeared.[7-9]

In invertebrates, opioid binding sites have been shown to be present in several distinct classes.[10-12] After the first demonstrations of endogenous opioids and relative binding sites in annellids and mollusks, several researchers pointed to other invertebrate groups, giving evidence of a widespread presence of an opioid system in all the phyla that were investigated, with the important exception of cnidaria. In fact, in *Hydra*, besides to other non-opioid peptides, only FMRFamide has been found, a peptide partially similar to enkephalins in the aminoacid sequence but functionally not related to opioids.[13] However, an effect of nalorphine has been reported on hydra GSH-induced feeding behaviour, thus suggesting that an opioid-like binding site may be present in this simple invertebrate.[14]

It is important to consider that some negative or contradictory results obtained in peptides or binding sites research in lower invertebrates may be due to a seasonal variability both in peptide levels[15] and in stereospecific opioid binding or to an incomplete identity in amino acid sequence between vertebrate and invertebrate molecules.[16]

As far as physiological roles are concerned, opioids in invertebrates seem to be involved in several important functions including neuromodulation, transmitter release (including nitric oxide), thermal behaviour, motility, gill ciliary activity, immunocyte activity and motility, digestion, sexual function, growth, vision, nociception and intraspecies aggression. For a review on opiates in invertebrates see reference 16.

The evidence that endogenous opioids and related receptors are present in invertebrates expanded their traditional role as vertebrate peptides. However, the early origin of this system during the phylogenesis is further demonstrated by studies indicating the presence in the prokaryote *Staphylococcus aureus* of Met-Enkephalin and of an opioid binding site. In this prokaryote, enkephalin decreased the growth rate and the incorporation of [^{3}H] thymidine. Naltrexone blocked this effect and, on its own, increased the replication rate, thus suggesting a tonic enkephalinergic inhibition of growth.[17,18]

Unicellular eukaryotes also seem to possess endogenous opioids and relative binding sites. β-endorphin has been identified in *Tethrahymena* and a saturable, reversible binding of labelled β-endorphin has been shown.[19,20] The opioid receptor found in this unicellular organism seems to lack the selectivity typical of higher forms, being able to bind ligands for all vertebrate opioid receptor subtypes. Opioids have been shown to inhibit pinocytosis and phagocytosis in *Amoeba* and in *Tethrahymena* and this effect is blocked by naloxone.[21,22]

Opioids in Planaria

Despite the fact that a direct demonstration of opioid receptors in planaria is lacking, behavioural studies support the presence of specific binding sites. The presence of opioid receptors in Planaria (*Dugesia gonocephala*) was first suggested by demonstrating a pharmacological action of morphine in this flatworm.[23] Morphine treatment has been reported to induce a reduction of spontaneous motor activity and of cAMP levels and this effect was antagonized by naloxone. Moreover naloxone, on its own, increased motor activity and its action has been suggested to involve dopamine release. There is clear evidence that stimulation of opioid receptors in vertebrates affects dopamine neurotransmission.[24] The interactions between opioids and the dopaminergic system have been widely studied in mollusks. The results obtained by Stefano et al[25,26] in *Mytilus edulis* and *Octopus bimaculatus* strengthen the concept that the activity of dopaminergic neurons may be modulated by afferent opioid signals and that, even in invertebrates, interneuronal transfer of information is more complex than formerly realized. Furthermore, the data confirm that endogenous opioids may exert a tonic control over dopamine metabolism, thus implying interdependence of the two systems.

Studies performed in planaria suggest that the removal by naloxone of the natural ligands from the specific receptors, by interrupting the enkephalinergic inhibitory modulation, may induce the dopamine release responsible for the hyperkinesias and the increase of cAMP levels. On the contrary, morphine induces a decrease in cAMP levels and a reduction of motor activity, due to an inhibition of dopamine release, as found also in *Mytilus edulis.*[26] The observation that naloxone-treated specimens show an increase of dopamine content, as demonstrated by neuronal fluorescence and by HPLC assay supports an enhanced release at the presynaptic level. This mechanism of action is further confirmed by the observation that the hyperkinetic patterns induced by naloxone are prevented in planaria specimens pretreated with reserpine or with haloperidol.[23] If the direct activity of naloxone on drug-naïve planaria would be confirmed, then a tonic action of enkephalins on dopamine release could be suggested.

The presence of endogenous opioids was directly demonstrated by means of radioimmunological and immunocytochemical methods in *Dugesia gonocephala*, using an anti-Met-enkephalin serum.[15] The Met-enkephalin-like immunoreactivity was localized in neurons and neuropil. Opioid levels in planaria show a significant seasonal variability, with lower levels in winter.

Opioid Receptor Subtypes in Planaria

More recent studies on the behavioural response of planaria exposed to selective opioid agonists demonstrate that, whereas both a μ agonist and a δ agonist failed to alter motor activity at all doses tested, the selective κ agonists U50,488 ((±)-trans-U-50-trans-3,4-dichloro-N-methyl-[2-(1-pyrrodinyl)-cyclohexyl]benzene acetamide methasulphonate) and bremazocine-HCl increased motor activity leading to "C-like position" (CLP) and "screw-like hyperkinesias" (SLH).[27] These changes were identical to those described under treatment with D2 or D1 dopamine receptor agonists, respectively.[28,29] Higher doses of κ agonists produced the enhancement of CLP and SLH together with robust movements defined as snake-like movements (SLM). This latter response, that was considered by the authors typical of stimulation of κ opioid receptors, was blocked by co-exposure to naloxone or the selective κ antagonist Nor-binaltorphimine (Nor-BNI).[27] The CLP response was inhibited by the D2 antagonist sulpiride whereas SLH hyperkinesias were inhibited by the D1 antagonist SH-23390. These data are in favour of the presence in planaria of opioid receptors similar to the κ subtype of mammals and indicate a functional interaction between the opioid and dopaminergic system in this simple animal model.[27] The differences of the behavioural effects of opioid agonists reported in this paper with respect to those previously described in the same species[23] could be explained by seasonal variability or by different experimental conditions.

Opioid Dependence in Planaria

Abstinence-induced withdrawal phenomena are behavioural manifestations of the physical dependence that develops to drugs of abuse. In mammals, withdrawal response is commonly used to assess and study dependence. As far as invertebrate models are concerned, a peculiar withdrawal phenomenon has been described in insects (crickets hyper-responsiveness to vibration).[30]

The consideration that planarians represent the most primitive and simple example of centralization of neuronal cells and possess a neurotransmitter system similar to the mammalian one, suggests that this flatworm may be particularly suitable as a model for dependence studies. Moreover, planarians respond with characteristic behavioural patterns following exposure to drugs acting on neural transmission.[28,29,31]

Using a quantitative method previously developed and tested to identify and quantify the cocaine abstinence induced-withdrawal behaviour in this flatworm, Raffa et al[32] demonstrated that following exposure to the κ-opioid agonist U-50,488H (trans-(±)-3,4-dichloro-N-methyl-N-(2-[1-pyrrolidinyl]cyclohexyl)-benzeneacetamide) planarians develop a physical dependence.

In fact, opioid-experienced specimens, when transferred into opioid-free water, show an acute abstinence-induced withdrawal phenomenon, consisting of a decrease in locomotor activity. The withdrawal phenomenon is precipitated by the opioid antagonist naloxone and the effect of U-50,488H is antagonized by a co-exposure to the non selective opioid antagonist naloxone or to the κ-selective antagonist nor-BNI. These findings further suggest the presence of κ-opioid receptors in planaria and offer the first direct neuropharmacological evidence of a receptor-mediated acute opioid withdrawal phenomenon in the flatworm.

Further studies using the same quantitative approach to analyse the competition between the κ-opioid agonist U-50,488H and the selective κ-antagonist nor-BNI confirmed these results and supplied evidence for an interaction of the two drugs at a common receptor, further supporting the view that a putative planaria opioid receptor might be functionally equivalent to the mammalian κ-opioid receptor.[33]

Cannabinoids and Opioids in Planaria

There is compelling evidence of similarities and interactions between opioid and cannabinoid systems in mammals, with particular reference to analgesia and drug addiction.[34] An endogenous cannabinoid system appears early along phylogeny[35] and the presence of an endogenous cannabinoid system in planaria has been reported by Paris and Lenicque.[36]

In planaria, exposure to the synthetic cannabinoid receptor agonist WIN 55212-2 induces dose-dependent stimulation of spontaneous motor behaviour, without stereotyped hyperkinesias.[37] High doses of the drug (250 µg/3 ml) induce stereotyped activities identical to those observed under treatment with opioid agonists, such as C-like position (CLP), screw-like hyperkinesias (SLH) and snake-like movements (SLM).[27] These effects are antagonized by co-exposure to a cannabinoid receptor antagonist (SR 141716A) or to the opioid receptor antagonist naloxone. These results indicate that functional interactions between cannabinoid and opioid systems are conserved along phylogeny.

Although there is no direct evidence of the presence of cannabinoid receptors in planaria, the dose-dependent effects of WIN 55212-2 and the dose-related antagonism by SR 141716A, together with the presence of specific cannabinoid receptor sites in low invertebrates such as cnidaria,[35] suggest that the behavioural changes observed may result from specific interactions of these drugs with a cannabinoid receptor.

The pattern of the hyperkinesias induced by the highest dose of WIN 55212-2 are very similar to those described by the same research group following exposure of planaria to opioid agonists and to dopaminergic agonists.[27-29] Those behavioural homologies were interpreted in favour of the hypothesis that the effects of WIN 55212.2 on motor activity in planaria might depend upon indirect stimulation of the endogenous opioid system of the flatworm, in turn influencing the control of dopamine release at the presynaptic level.[37]

Poly-drug exposure was also studied in planaria[38,39] by measuring the effects of a co-exposure to opioids and cocaine on withdrawal behaviour. The results demonstrated a subadditive interaction of the two drugs: the co-exposure to cocaine and a κ-opioid agonist attenuates the development of dependence and subsequent abstinence-induced withdrawal. Of interest in this regard are the observations[40] suggesting that in planaria cocaine may inhibit dopamine reuptake and/or stimulate dopamine release, thus inducing an indirect dopaminergic action, also confirmed by changes in cAMP levels.

In planaria, the same effects on dopaminergic transmission are induced by κ-opioid agonists.[27] This common activity of opioids and cocaine could help to explain the subadditive interaction between cocaine and U-50,488H for abstinence-induced withdrawal.

Co-exposure of the planarians to D-glucose or to 2-deoxy-D-glucose (but not to L-glucose) attenuates the development of physical dependence, shown by an attenuated withdrawal syndrome, from cocaine and U-50,488H.[39] These results suggest that either D-glucose and 2-deoxy-D-glucose compete with a common cocaine and κ-opioid transport mechanism or that the development of physical dependence (or the inhibition of abstinence-induced withdrawal) in planarians requires energy supplied from glucose metabolism.

Endogenous Opioid Alkaloids in Flatworms

The analysis of the different subtypes of opioid receptors revealed that a particular μ receptor, classified as $μ_3$, binds only opioid alkaloids and is insensitive to opioid peptides. The physiological function of this receptor, that was found in the immunocytes of the mussel *Mytilus edulis* as in human granulocytes and monocytes and in other cells such as those of the nervous system,[41-43] was unclear until the demonstration that several animal cells, from both vertebrates and invertebrates, are able to synthesize the alkaloid morphine and other opioid alkaloids such as morphine-6-glucoronide and codeine.[44] The stimulation of $μ_3$ receptors induces several responses, including the down regulation of the immune system.

Of particular interest is the demonstration that a parasitic flatworm, *Schistosoma mansoni* as well as other parasitic helmintes such as *Ascaris suum* contain endogenous morphine. A morphine-like molecule was also detected in *Schistosoma mansoni* using HPLC and radioimmune assay (RIA).[45] Recently, these data have been confirmed by more specific methods, using both HPLC and MS.

Morphine has been reported to be immunosuppressive. This ability of morphine to suppress the immune response of the host has led to the hypothesis of the secretion of morphine by parasitic invertebrates for this purpose.

The endogenous material detected in *Schistosoma*, corresponding to morphine, was found to mimic authentic morphine in its ability to induce immunocyte rounding and immobility, an action that is naloxone sensitive.[45,46] In addition, morphine levels in *Schistosoma* are enhanced after coincubation with human polymorphonuclear leukocytes, a situation mimicking the exposure to the host immune system. This alkaloid could play a role both as an internal signal in the worm, possibly by regulating NO release and as an external signal with a possible immune suppressive function. Indeed, it is well known that several parasites, including *Schistosoma* and *Ascaris* are virtually invisible to the immune system of the hosts, suggesting that the parasite may utilize morphine to escape host immune response. It is interesting to note that *Schistosoma* is invisible not only to the immune system of the human host but also to the intermediate snail host. Only living parasites can escape the immune response and if the parasite dies then an acute immune response is activated by the hosts.

Conclusions and Perspectives

Several types of evidence are necessary to support the theory that an opioid system is present in a living species: (i) the endogenous ligands (peptides and/or alkaloids) and the genes encoding for their synthesis should be demonstrable; (ii) stereospecific receptors and related genes should be present; (iii) the organisms should display a measurable and reversible response to the opioids.

Regarding planaria, even if several evidences taken together support the presence of an opioid system (immunochemical demonstration of met-enkephalin, pharmacological evidence of a measurable and reversible response to opioids, abstinence-induced withdrawal phenomena), the data are at present incomplete and should be supplemented with genetic data, stereospecific binding studies and more conclusive demonstration of endogenous ligands, supported by direct biochemical analysis.

The presence of the endogenous morphine alkaloid in the flatworm *Schistosoma mansoni* strongly suggests that the possible presence of opioid alkaloids in planaria and their role as internal signals should be investigated. Of great interest moreover would be the search for $μ_3$-like receptors in flatworms, as a possible primitive sub-type of opioid binding site.

A large body of research performed over the last decades suggests a strong conservation of basic neurochemical mechanisms during evolution. Although the nervous system of planaria is extremely simple compared with the vertebrate nervous system, it seems to represent a useful model for the study not only of the evolution of neurotransmission, but also of the action mechanism of opioids and of abstinence-induced withdrawal phenomena.

References

1. Pert CB, Snyder SH. Opiate receptor: demonstration in nervous tissue. Science 1973; 179:1011-1014.
2. Hughes J. Isolation of an endogenous compound from the brain with pharmacological properties similar to morphine. Brain Res 1975; 88:295-308.
3. Goldstein A. Naidu A. Multiple opioid receptors: ligand selectivity profiles and binding site signatures. Mol Pharmacol 1989; 36:265-272.
4. Mansour A, Khachaturian H, Lewis ME et al. Autoradiographic differentiation of mu, delta and kappa opioid receptors in the rat forebrain and midbrain. J Neurosci 1987; 7:2445-2464.
5. Pert CB, Aposhian D, Snyder SH. Phylogenetic distribution of opiate receptor binding. Brain Res 1974; 75:356-361.
6. Simantov R, Goodman R, Aposhian D et al. Phylogenetic distribution of a morphine-like peptide 'enkephalin'. Brain Res 1976; 111:204-211.
7. Remy C, Dubois MP. Localisation par immunofluorescence de peptides analogues a l'α-endorphine dans le ganglions intra-aesophagiens du lombricide Dendrobaena subrubiconda Eisen. Experientia 1978; 35:137-138.
8. Alumets J, Hakanson R, Sundler F et al. Neuronal localisation of immunoreactive enkephalin and beta-endorphin in the earthworm. Nature 1979; 279:805-806.
9. Stefano GB, Kream RM, Zukin RS. Demonstration of stereospecific opiate binding in the nervous tissue of the marine mollusc Mytilus edulis. Brain Res 1980; 181:440-445.
10. Kream RM, Zukin RS, Stefano GB. Demonstration of two classes of opiate binding sites in the nervous tissue of the marine mollusc Mytilus edulis. Positive homotropic cooperativity of lower affinity binding sites. J Biol Chem 1980; 255:9218-9524.
11. Kavaliers M, Rangeley RW, Hirst M et al. Mu- and kappa-opiate agonists modulate ingestive behaviors in the slug, Limax maximus. Pharmacol Biochem Behav 1986; 24:561-6.
12. Santoro C, Hall LM, Zukin RS. Characterization of two classes of opioid binding sites in Drosophila melanogaster head membranes. J Neurochem 1990; 54:164-170.
13. Grimmelikhuijzen CJP, Dockray GJ, Schot LPC. FMRFamide-like immunoreactivity in the nervous system of Hydra. Histochemistry 1982; 74:499-508.
14. Venturini G, Carolei A, Palladini G et al. Peptide-monoamine interactions in Planaria and hydra. In: Stefano GB, ed. Handbook of comparative opioid and related neuropeptide mechanisms. Boca Raton: CRC Press, 1986: 245-254.
15. Venturini G, Carolei A, Palladini G et al. Radioimmunological and immunocytochemical demonstration of Met-enkephalin in Planaria. Comp Biochem Physiol 1983; 74C:23-25.
16. Harrison LA, Kastin AJ, Weber JT et al. The opiate system in invertebrates. Peptides 1994; 15:1309-1329.
17. Zagon IS, Goodman SR, McLaughlin PJ. Characterization of zeta (zeta): a new opioid receptor involved in growth. Brain Res 1989; 482:297-305.
18. Zagon IS, McLaughlin PJ. An opioid growth factor regulates the replication of microorganisms. Life Sci 1992; 50:1179-1187.
19. LeRoith D, Shiloach J, Roth J et al. Evolutionary origins of vertebrate hormones: material very similar to adrenocorticotropic hormone, beta-endorphin and dynorphin in protozoa. Trans Assoc Am Physicians 1981; 94:52-60.
20. O'Neill JB, Pert CB, Ruff MR et al. Identification and characterization of the opiate receptor in the ciliated protozoan, Tetrahymena. Brain Res 1988; 450:303-315.
21. Josefsson JO, Johansson P. Naloxone-reversible effect of opioids on pinocytosis in Amoeba proteus. Nature 1979; 282:78-80.
22. DeJesus S, Renaud FL. Phagocytosis in Tetrahymena termophila: Naloxone-reversible inhibition by opiates. Comp Biochem Physiol 1989; 92C:1139-1142.
23. Venturini G, Carolei A, Palladini G et al. Naloxone enhances cAMP levels in Planaria. Comp Biochem Physiol 1981; 69C:105-108.
24. Di Chiara G, Imperato A. Drugs abused by humans preferentially increase synaptic dopamine concentrations in the mesolimbic system of freely moving rats. Proc Natl Acad Sci USA 1988; 85:5274-5278.
25. Stefano GB. Comparative aspects of opioid-dopamine interaction. Cell Mol Neurobiol 1982; 2:167-178.
26. Stefano GB, Hall B, Makman MH et al. Opioid inhibition of dopamine release from nervous tissue of Mytilus edulis. Science 1981; 213:928-930.
27. Passarelli F, Merante A, Pontieri FE et al. Opioid-dopamine interaction in planaria: a behavioral study. Comp Biochem Physiol 1999; 124C:51-5.
28. Carolei A, Margotta V, Palladini G. Proposal of a new model with dopaminergic-cholinergic interactions for neuropharmacological investigations. Neuropsychobiology 1975; 1:355-364.

29. Venturini G, Stocchi F, Margotta V et al. A pharmacological study of dopaminergic receptors in Planaria. Neuropharmacology 1989; 28:1377-1382.
30. Zabala NA, Gomez MA. Morphine analgesia, tolerance addiction in the cricket Pteronemobius sp. (Orthoptera, Insecta). Pharmacol Biochem Behav 1991; 40:887-891.
31. Palladini G, Ruggeri S, Stocchi F et al. A pharmacological study of cocaine activity in Planaria. Comp Biochem Physiol 1996; 115C:41-45.
32. Raffa RB, Stagliano GW, Umeda S. κ-Opioid withdrawal in Planaria. Neurosci Lett 2003; 349:139-142.
33. Raffa RB, Baron DA, Tallarida RJ. Schild (apparent pA2) analysis of a κ-opioid antagonist in Planaria. Europ J Pharmacol 2006; 540:200-201.
34. Manzanares J, Corchero J, Romero J et al. Pharmacological and biochemical interactions between opioids and cannabinoids. Trends Pharmacol Sci 1999; 20: 287-294.
35. De Petrocellis L, Melck D, Bisogno T et al. Finding of the endocannabinoid signalling system in Hydra, a very primitive organism: possible role in the feeding response. Neuroscience 1999; 92:377-387.
36. Paris M, Lenicque PM. Effect of tetrahydrocannabinol and of cannabidiol on wound healing and regeneration of the Planaria Dugesia tigrina. Therapie 1975; 30:97-102.
37. Buttarelli FR, Pontieri FE, Margotta V et al. Cannabinoid-induced stimulation of motor activity in Planaria through an opioid receptor-mediated mechanism. Prog Neuropsychopharmacol Biol Psychiatry 2002; 26:65-68.
38. Raffa RB, Stagliano GW, Tallarida RJ. Subadditive withdrawal from cocaine κ-opioid agonist combinations in Planaria. Brain Research 2006; 1114:31-35.
39. Umeda S, Stagliano GW, Raffa RB. Cocaine and κ-opioid withdrawal in Planaria blocked by D-, but not L-, glucose. Brain Research 2004; 1018:181-185.
40. Palladini G, Ruggeri S, Stocchi F et al. A pharmacological study of cocaine activity in Planaria. Comp Biochem Physiol 1996; 115C:41-45.
41. Stefano GB, Digenis A, Spector S et al. Opiate-like substances in an invertebrate, an opiate receptor on invertebrate and human immunocytes and a role in immunosuppression. Proc Natl Acad Sci USA 1993; 90:11099-11103.
42. Makman MH, Bilfinger TV, Stefano GB. Human granulocytes contain an opiate alkaloid-selective receptor mediating inhibition of cytokine-induced activation and chemotaxis. J Immunol 1995; 154:1323-1330.
43. Stefano GB, Hartman A, Bilfinger TV et al. Presence of the mu3 opiate receptor in endothelial cells. Coupling to nitric oxide production and vasodilation. J Biol Chem 1995; 270:30290-3.
44. Stefano GB, Scharrer B. Endogenous morphine and related opiates, a new class of chemical messengers. Adv Neuroimmunol 1994; 4:57-68.
45. Pryor SC, Henry S, Sarfo J. Endogenous morphine and parasitic helminthes. Med Sci Monit 2005; 11: RA183-189.
46. Salzet M, Capron A, Stefano GB. Molecular crosstalk in host-parasite relationships: schistosome- and leech-host interactions. Parasitology Today 2000; 16:536-40.

CHAPTER 6

Nitric Oxide in Lower Invertebrates

Marco Colasanti* and Giorgio Venturini

Abstract

Nitric oxide (NO) is considered an important signaling molecule involved in many different physiological processes, including nervous transmission, vascular regulation, immune defense and in the pathogenesis of several diseases. The presence and roles of NO are well demonstrated in the main invertebrate groups, showing the widespread presence of this signaling molecule throughout the animal kingdom, from porifera up to higher invertebrates. Conscious of the restrictions involved in the choice they made, the authors attempt to provide a survey of current knowledge of the genesis and possible roles of NO and the related signaling pathway in lower invertebrates (i.e., porifera, cnidaria and platyhelmintes). Unfortunately, the great diversity of nitric oxide synthase (NOS) family within the animal kingdom clearly prevents interpretation of the emerging comparative data in terms of the existing classification of mammalian NOS isoforms. Nevertheless, all of these results indicate that NO is one of the earliest and most widespread signaling molecules in living organisms.

Introduction

Nitric oxide (NO) is considered an important signaling molecule involved in diverse physiological processes, including nervous transmission, vascular regulation, immune defense and in the pathogenesis of several diseases. NO is an unstable nitrogen radical spontaneously degrading into nitrites and is generated by the conversion of L-arginine into L-citrulline, through the NO synthase (NOS) enzyme.[1] In mammals, two enzymes, the neuronal (nNOS or NOS-I) and the endothelial (eNOS or NOS-III) isoforms, are Ca^{2+}/calmodulin-dependent and constitutively expressed (also termed constitutive NOS or cNOS). In turn, NO can exert its biological activity via the stimulation of soluble guanylate cyclase,[2] thus leading to an increase in cGMP, or by interacting with a number of other enzymes/proteins. The third enzyme is an inducible Ca^{2+}-independent isoform (iNOS or NOS-II), expressed in some cell types after stimulation with Escherichia coli lipopolysaccharide (LPS) and/or different cytokines such as interferon-γ (IFNγ), interleukin-1β (IL-1β), or tumor necrosis factor-α (TNFα).[3]

The presence and roles of NO are well demonstrated in the main invertebrate groups, showing the widespread occurrence of this signaling molecule throughout the animal kingdom, from higher invertebrates down to porifera and even to prokaryotic cells. In invertebrates, experimental evidence suggests the presence of NOS isoforms other than those known for higher organisms. Noteworthy is the early appearance of NO during evolution and striking is the role played by the nitrergic pathway in the sensorial functions, from coelenterates up to mammals, mainly in olfactory-like systems.

The increasing amount of scientific literature regarding NO in invertebrates prevents the authors from providing a widespread coverage of the topic. Therefore, the present chapter

*Corresponding Author: Marco Colasanti—Dipartimento di Biologia, Università di Roma Tre, Viale Marconi 446, 00146 Roma, Italy. Email: colasant@uniroma3.it

Planaria: A Model for Drug Action and Abuse, edited by Robert B. Raffa and Scott M. Rawls.
©2008 Landes Bioscience.

focuses its attention on some aspects of the biological role of NO in lower invertebrates (i.e., porifera, cnidaria and platyhelmintes), integrating the extensive reviews already available on this topic.[4-8] Certainly, the authors are conscious of the restrictions imposed by this option, thereby inviting the interested reader to consult the many valuable papers on the NO pathway in higher invertebrates.[4-10] All literature data here reported suggest that future research on the biological roles of early molecules in lower living forms is likely to be very important for the understanding of the evolution of signaling systems.

Nitric Oxide in Porifera

Sponges are the oldest metazoan group, living in the sea ever since the Precambrian period (600 million years ago). The first clear evidence of the presence of NO pathway in sponges was provided in 2001.[11] In that study, NOS activity was demonstrated by evaluating the conversion of radioactive citrulline from [^{14}C]arginine in intact cells into two different species, *Axinella polypoides* and *Petrosia ficiformis*, which are phylogenetically unrelated within the class of Demospongiae. NO production was also confirmed by an electron paramagnetic resonance analysis. Moreover, the histochemistry technique of NADPH diaphorase showed a specific localization of NOS activity in a particular network of dendritic cells in the sponge parenchyma.[11] NOS enzymatic activity resulted to be calcium dependent, strongly suggesting an analogy with the constitutive isoform of other invertebrates, where NOS activity is involved mainly in cellular signaling in response to sensorial stimuli. Note that NOS activity was measured at 18°C on intact cells obtained by mechanical dissociation of the sponges. Interestingly, exposure of cells to heat stress (i.e., 28°C for 3 h) increased NOS activity in both *Axinella polypoides* and *Petrosia ficiformis* cells, thus suggesting that NO may have evolved in Porifera as an ancient cellular signal of environmental stress.[11]

The question now is how heat stress increases NOS activity. Previous results seem to indicate that temperature can directly influence NOS activity independently of the substrate concentration, the values of Km being temperature independent.[12] However, an indirect effect (e.g., increase of intracellular Ca^{2+}) may be also conceivable. In this respect, it has been demonstrated that sponges express ADP-ribosyl cyclase activity, which converts NAD^+ into cyclic ADP-ribose, a potent and universal intracellular Ca^{2+} mobilizer.[13] In *Axinella polypoides*, ADP-ribosyl cyclase was activated by temperature increases by means of an abscisic acid-induced, protein kinase A-dependent mechanism. The thermosensor triggering this signaling cascade was a heat-activated cation channel.[13]

Soon afterwards, several studies were carried out on sponges, suggesting that the biochemical pathway leading to NO formation is widely distributed in this group. Recently, NO production has been demonstrated in the demosponge *Suberites domuncula*.[14] In this study, the role of NO has been evaluated in the radiation-induced bystander effect, the latter being well described in vertebrates. In radiation-induced bystander effects, important biological effects of ionizing radiation arise in cells that receive no radiation exposure as a consequence of damage signals transmitted from neighboring irradiated cells; transmission may be mediated either by direct intercellular communication through gap junctions, or by factors released into the surrounding medium. In this phenomenon, the biological effects appear to be associated with an upregulation of oxidative metabolism. For their experiments,[14] the authors used sponges in the two-chamber-system. The lower dish contained irradiated "donor" cells (single cells) and the upper dish the primmorphs ("recipient" primmorphs). Primmorphs are very densely packed spherical-shaped sponge-cell aggregates with a continuous pinacoderm (skin cell layer) covered by a smooth, cuticle-like structure. The "donor" cells were treated with UV-B light and hydrogen peroxide (H_2O_2), these factors being present also in the natural marine aquatic environment of sponges; these factors caused a high level of DNA strand breaks followed by a reduced viability of the cells. If these cells were added to the "recipient" primmorphs, these 3D-cell cultures started to undergo apoptosis.[14] Other experiments with *Suberites domuncula* revealed that (i) the bystander effect was controlled by NO and (ii) the major enzyme modulating this effect was dimethylarginine dimethylaminohydrolase (DDAH). DDAHs metabolize methylated arginines, e.g., asymmetric dimethylarginine (ADMA) to citrulline. ADMA, a competitive NOS inhibitor, reduces the production of NO.[15]

Based on the experimental data, the following sequence of events can be considered to be the major elements of the bystander response in the sponge system. In response to UV irradiation and H_2O_2 and in the absence of ethylene, the concentration of ADMA is low, due to the high level of DDAH which causes a hydrolysis of this physiological NOS inhibitor. If the "donor" cells are exposed to ethylene the level of DDAH expression becomes downregulated, as a consequence causing the ADMA concentration to increase, while resulting in a decreased of NO release and a lower deleterious bystander effect.[14] It can be deduced that in sponges the bystander response is involved in the host defense mechanisms both against xenobiotica and in attacking microorganisms in their environment.[14]

Despite the fact that sponges are nerveless and muscleless multicellular animals, they are able to react upon external stimuli, to move, to contract and to display circadian rhythms. The earliest description of such sponge behavior dates back to Aristotle, who mentioned contraction of living sponges, when they were touched and collected by humans for the production of bath sponges. Recently, NO has been found to be involved in the coordination mechanisms of the contractile and locomotive demosponge *Tethya wilhelma*.[16] By using digital time lapse imaging of the sponge body, in conjunction with quantitative image analysis, the authors observed that the NO-releasing substance NOC-12 was able to induce contractions instantly after application. The contractions, which occurred within the time of exposure to NO, were similar to non-induced endogenous contractions, except for the attenuated amplitude: they seemed to be triggered locally and spread over the sponge body in a wave-like manner. Due to its fast diffusion, the short half life and the limited range, NO seems to be an optimal candidate for the regulation and timing of the endogenous contraction rhythm within pacemaker cells in *Tethya wilhelma*.[16]

Thus, the existence of coordination mechanisms, possibly based on the release of NO in the sponge matrix, represents a cell signaling system that may be considered the most primitive sensorial network in the animal kingdom.

Nitric Oxide in Cnidaria

Increasing evidence supports the existence of NO pathway also in cnidaria. In 1995, the presence of NO pathway was first evidenced in the freshwater coelenterate *Hydra*, one of the most primitive organism possessing a nervous system.[17] Soon afterwards, further evidence was obtained, suggesting that the NO signaling system is widespread in this phylum. In particular, positive NADPH diaphorase staining has been reported in a number of organisms including the following: the cnidarian *Aiptasia diaphana* in supporting cells surrounding the nematocytes;[18] the marine cnidarian *Aiptasia pallida* in the epidermal cells and at the extremities of the mesoglea;[19,20] the jellyfish *Aglantha digitale* in neurites running in the outer nerve ring at the base of the animal and in putative sensory cells in the ectoderm covering its tentacles;[21] the sea pansy *Renilla koellikeri*, an octocorallian of the sea pen family, in the basiectoderm at the base of the autozooid polyp tentacles and in a nerve-net around the oral disc.[22]

Parallel to NADPH diaphorase staining, the enzymatic activity, as verified by evaluating the conversion of [3H]arginine to [3H]citrulline, was also used to detect NOS in cnidaria. In this respect, biochemical studies have demonstrated that *Hydra vulgaris* constitutively express a NADPH/Ca^{2+}-dependent (but calmodulin-independent) NOS activity.[23] Further experiments have shown that *Hydra* are able to release basal levels of NO, as determined by measuring nitrite, the NO breakdown product. Nitrite production was also verified by optical and ESR spectroscopy.[23] Moreover, in *Aiptasia diaphana* a Ca^{2+}-dependent citrulline-forming enzymatic activity was found in the acontial tissue[18] and the release of NO during the discharge of in situ nematocytes was assessed by optically monitoring the NO-induced oxidation of oxyhaemoglobin to methaemoglobyn.[18] Again, high levels of nitrite and nitrate were detected in the tentacles of the jellyfish *Aglantha digitale*.[21] Furthermore, NOS activity was well characterized in *Aiptasia pallida*.[19] The NADPH-dependent NOS activity was predominantly cytosolic and was characterized by a Km for arginine of 19.05 mM and a Vmax of 2.96 pmol/min per µg protein.[19] In addition, NOS activity was measured in a variety of coral species, including *Madracis mirabilis*, *Madracis decactis*, *Agaricia sp.* and *Montastrea franksi.* The mean activity in corals as well as in anemones was similar, typically in the range 0.50-1.16 pmol/min per μg protein. Finally, the use of the recently available NO sensitive fluorescent dye DAF-FM and laser scanning confocal microscopy have enabled detection of NO in vivo.[20] Interestingly, these authors have identified a LPS-inducible NOS activity (e.g., NOS-II-like) in *Aiptasia pallida.*[20] The presence of an inducible NOS activity in *Aiptasia pallida* was also investigated by Morral et al.[19] In this study, however, pretreatment with LPS, either by injection of 20 μg directly into the oral cavity or by immersion for up to 24 h in a solution of 2 μg/ml, caused no discernible alteration in NOS activity of the whole homogenate.[19] Very recently, NOS from cnidarian (*Discosoma striata*) has been found to lack a distinct structural element that is present as an insertion in the reductase domains of constitutive NOSs but it is absent in inducible NOSs of vertebrates.[24] This insert of about 45 amino acids is thought to be an autoinhibitory loop which impedes Ca^{2+}/calmodulin binding and enzymatic activation. Interestingly, since *Discosoma striata* NOS is structurally similar to both the only one known nonanimal conventional NOS and vertebrate inducible NOS, it is tempting to speculate that the inducible type of NOS can be evolutionarily ancestral for animal NOSs.[24]

Biological roles of NO in cnidaria are multiple and are related to feeding, defense, environmental stress, swimming, symbiosis and peristaltic muscle activity. In *Aiptasia diaphana*, NO is involved in the discharge of acontial nematocytes, specialized stinging cells present in the tentacles which have a variety of functions, most usually in defense or capture of prey species.[18] When exposed to an appropriate stimulus, these cells release a fluid in which various toxins are stored. The triggering of discharge requires stimulation both of chemoreceptive sites on supporting cells and of the ciliated mechanoreceptor on the nematocyte. The chemical stimulus induces calcium influx into supporting (or sensory) cells that activates NOS. The NO released reaches the resting nematocyte where it increases its excitability to mechanical stimuli applied to the ciliary receptor so inducing the discharge.[18]

NO has been found to play an important role in the induction and control of feeding response in *Hydra vulgaris.*[17,23] Hydra is a sessile predator whose tentacles are armed with the typical, characteristic stinging capsules of the coelenterates called nematocysts. When a prey accidentally touches a tentacle, a typical feeding response, consisting of tentacle writhing and mouth opening, is activated. The reduced glutathione (GSH) outflow from the prey when pierced by tentacle nematocysts is the physiological activator of hydra feeding response. Interestingly, NO participates in the triggering and coordination of tentacular curling, providing for the fast diffusion of a primary stimulus to the neighboring tentacles, regardless of any direct connection through synapses (tentacle recruitment).[23] Initially, GSH induces the feeding response through mediators such as cAMP, IP_3 and/or Ca^{2+}, while successively, an increase of GSH-induced Ca^{2+} levels is responsible for NO production, which in turn elicits the recruitment of neighboring tentacles. Note that NO stimulus alone is not sufficient for complete induction of the complex behavioral phenomenon. Finally, on a longer timescale, elevated NO-induced cGMP levels are able to trigger inhibition of the GSH-induced feeding response.[17] These mechanisms are consistent with those reported for the mammalian olfactory system.[25]

NO/cGMP signaling pathway regulates the swimming program in the jellyfish *Aglantha digitale.*[21] This cnidarian exhibits two main forms of swimming: one, fast, related to escape and another one, slow, related to fishing. NO endogenously produced in putative nitrergic sensory neurons in the tentacles acts as a neuromodulatory agent in controlling the rhythmic slow swimming associated with feeding.[21] In the sea pansy *Renilla koellikeri*, NO/cGMP pathway plays a pivotal role in peristalsis, an activity that moves body fluids through the coelenteron (gastrovascular cavity) of the polyps across the colony.[22] In this system, NO donors increase the amplitude of peristaltic contractions and increase tonic contractions in relaxed preparations, but cause a relaxation of basal tension in contracted preparations.[22] The analogy with the NO-mediated modulation of vascular tone and of peristalsis in mammals suggests that these physiological roles of NO were conserved throughout the evolution of metazoans. While other roles of NO were proposed for cnidarians, such as in the feeding behavior of *Hydra*,[17,23] the nematocyst discharge of a sea anemone[18] and the swimming behavior of a jellyfish,[21] the present is the first report of NO involvement in cnidarian activities related to internal fluid movement.

Recently, NO is emerging as an important regulator in both parasitic and mutualistic symbioses in cnidaria.[20] In this respect, the production of NO has been described as part of the cellular stress response of the symbiotic sea anemone *Aiptasia pallida*, which hosts dinoflagellates from the genus *Symbiodinium*. In particular, it has been observed that upon exposure to elevated temperatures, symbiotic anemones responded to algal-derived ROS (including superoxide and hydrogen peroxide) by producing high NO levels through a signaling pathway leading to the upregulation of NOS (e.g., NOS-II like).[20] The reaction of NO with superoxide produced the reactive nitrogen species peroxynitrite, with deleterious effects leading to cell death and to the collapse of the symbiosis, a detrimental process known as coral (cnidarian) bleaching. This response is similar to that described in some mammalian systems exposed to pathogens together with oxidative stress and therefore it has important evolutionary implications.

Nitric Oxide in Platyhelminthes

Although the first study conducted by Elofsson et al[5] failed to find a NO pathway in flatworms, in 1996 two groups reported the presence of NOS activity in two different species of platyhelminthes.[26,27] In particular, NADPH diaphorase staining was studied in the freshwater planarian *Dugesia tigrina*.[26] The positive staining was restricted to the pharynx, where the inner epithelium was intensely stained and the outer epithelium moderately stained. Neurons that innervated the pharynx were also stained. NOS enzyme activity was also confirmed by HPLC quantitation of the formed citrulline.[26] At the same time, the presence of NOS was detected in the nervous system of adult *Hymenolepis diminuta*.[27] Positive NADPH diaphorase staining was observed in nerve fibers in the main and minor nerve cords and the transverse ring commissures and in cell bodies in the brain commissure, along the main nerve cords, in the suckers and the rostellar sac. Positive staining was also observed in the wall of the internal seminal vesicle and the genital atrium.[27] More recently, NOS activity and the rate of NO release by the worm have been also detected measuring the conversion of L-[³H]arginine into L-[³H]citrulline[28] and the accumulation of nitrites and nitrates, using the Griess reaction,[29] respectively.

Soon afterwards, major evidence was obtained confirming that the NO pathway occurs also in other plathyhelmintes. In adult *Diphyllobothrium dendriticum*, NADPH-diaphorase staining was observed in neurons in the bilobed brain and along the 2 main nerve cords and in nerve fibers close to the body musculature and the musculature of the reproductive ducts, in the walls of the testicular follicles and in sensory endings in the tegument.[30] In the flatworm *Mesocestoides vogae tetrathyridia*, NADPH diaphorase staining occurred in the brain and the main nerve cords but also followed the muscle fibers.[31] The distribution of NADPH diaphorase reactivity was also analyzed in the adult fluke *Fasciola hepatica*.[32] Positive NADPH diaphorase staining occurred in neuronal tissue and in nonneuronal tissue. Large, NADPH diaphorase-stained neurones were localized in the nervous system. The oral and ventral suckers were innervated with many large NADPH diaphorase-stained neurones. In addition, the NADPH diaphorase staining reaction followed closely the muscle fibers in both the suckers, in the body and in the ducts of the reproductive organs.[32] The activity of neuronal NOS in homogenates of *Fasciola hepatica* was also measured by the direct radiometric assay of the production of L-[³H]citrulline.[33] With histochemical detection of NADPH diaphorase, the presence of NOS was demonstrated in the digenetic trematode, *Fasciolopsis buski*.[34,35] Strong NADPH diaphorase staining was observed in the neuronal cell bodies in the two cerebral ganglia, the brain commissure and the nerve fibers in the main nerve cords. Positive staining was also detectable in the innervation of the pharynx, the cirrus sac and the ventral sucker besides being observable sporadically in the nerve tributaries in the general parenchyma.[34,35] NO released by the whole worm kept in PBS at 37°C was also measured spectrophotometrically. The NOS activity was assayed in the whole worm homogenate and also in the tissue homogenate containing only the anterior pre-acetabular part of the parasite body.[34,35] The pattern of the NADPH diaphorase reaction was also investigated in cercaria of *Diplostomum chromatophorum*.[36] The NADPH diaphorase reaction occurred in the ventral sucker, the hind body and the tail.[36] Very recently, by using NADPH diaphorase staining, the existence of NOS activity has been demonstrated in *Taenia solium* metacestode, a larval pork tapeworm that causes systemic infections in humans.[37] Moreover, an NOS-I-like molecule of approximately 95 kDa was detected using antibodies against NOS-I N-terminus mapping either by immunoprecipitation or by Western blotting.[37] In adult *Schistosoma mansoni*, NOS immunoreactivity and NOS activity were analyzed using 3 different types of NOS antibodies (i.e., anti-NOS-I, anti-NOS-II and anti-NOS-III) and NADPH-diaphorase histochemistry.[38] NOS-I-like immunoreactivity was found in the main nerve cords and the peripheral nervous system. Putative sensory neurons with apical neuronal processes leading to the tegument of male worms were also immunoreactive for NOS-I. Anti-NOS-II labeled a variety of predominantly nonneuronal tissues, showing intense labeling at or near the surface of the worm and in components of the gastrointestinal tract. Anti-NOS-III showed no selective labeling. The distribution of NADPH-diaphorase reactivity was generally similar to the pattern of NOS immunoreactivity, including labeling of neuronal-like cells as well as developing eggs.[38] The presence of an inducible NOS isoform and its distribution have been further investigated in *Schistosoma japonicum* and *Schistosoma mansoni* in different stages of the parasites.[39] Reactivity was associated with the tegument in both larval schistosomes (sporocysts and cercariae) and eggs. In adult worms, the majority of the immunofluorescence was predominantly subtegumental in *Schistosoma japonicum* and parenchymal in *Schistosoma mansoni*. In Western blots, the enzyme of *Schistosoma japonicum* had an apparent molecular weight of about 210 kDa.[39] A direct estimate of the presence and distribution of authentic NO in living schistosomes was provided using NO sensitive fluorescent dye DAF-2 diacetate.[40] In adult worms, DAF-2 fluorescence was found selectively in epithelial-like cells.[40]

The biological roles of NO in platyhelminthes are multiple and are related to the presence and distribution of NOS isoforms in organisms. In general, NO seems to be involved in maintaining the physiological homeostasis (e.g., feeding, muscle activity, nitrergic signaling, reproduction and development). Interestingly, *Planaria* are considered as a useful model for studying drug withdrawal.[41,42] Recently, it has been observed that the NOS inhibitor L-NAME attenuates abstinence-induced withdrawal from both cocaine and a cannabinoid agonist, suggesting that the withdrawal from these drugs in planarians is mediated, at least in part, by NO pathway.[43] As already observed for *Hydra*, NO can play a role in some aspects of feeding behavior in *Planaria*, NOS activity being mainly localized in the pharynx.[26] Consistently with this role, the effects of a NO donor and a NOS inhibitor were studied on the sucker musculature of *Mesocestoides vogae tetrathyridia* by following the rate at which they attach to each other with the aid of their suckers, thus forming aggregates or clusters.[31] NO donor increased the number of clusters formed, whereas N-nitro-L-arginine, which is a NOS inhibitor, decreased the rate of cluster formation, indicating that NO is needed for the normal activity of the sucker musculature in *Mesocestoides vogae tetrathyridia*.[31] A possible role for NO in the control of muscle activity is also supported by the presence of NADPH diaphorase activity along muscle fibers in *Fasciola hepatica* as well as in other flatworms.[32] Furthermore, this role has been confirmed in adult *Hymenolepis diminuta* by studying the pattern of cGMP immunostaining after stimulation with a NO donor. cGMP was detected in the peripheral nervous system, especially in nerve fibers close to the body muscle fibers. cGMP also occurred in terminals beneath the basal lamina of the tegument and between the muscle fibers of the suckers.[44]

It is generally accepted that the platyhelminthes contains four major groups, each having a unique anatomy, body size and life style: the 'Turbellaria' (a paraphyletic assemblage of at least seven distinct lineages of mostly free-living forms), Monogenea (primarily ectoparasitic), Trematoda (endoparasitic flukes) and Cestoda (endoparasitic tapeworms). Among these, the latter three groups (called 'Neodermata') are represented by diverse obligate parasitic flatworms of invertebrates and vertebrates that cause diseases in a variety of host animal groups, including domestic animals and humans. Although the involvement of NO (with particular regard to inducible NOS isoform) in host response to infection by a variety of parasitic platyhelminthes has been widely investigated, there is little information available regarding the role, or even the presence, of an inducible NO pathway in parasite themselves. As described above, *Schistosoma japonicum* and *Schistosoma mansoni* were tested for reactivity with an anti-NOS-II antibody and the distribution of NOS-II was studied by immunofluorescent tests in different stages of the parasites. These results suggested that an NOS-II-like enzyme was present in schistosomes and indicated potential roles in reproduction and development.[38,39] However, the role of worm and host NOS-II in the parasite-host interrelation remains to be clarified. Very recently, in order to assay the physiological role and downstream targets of NO in adult schistosomes, changes in parasite gene expression in response to exposure to exogenous NO in vitro have been examined by using Long-SAGE (serial analysis of gene expression).[45] Finally, a role of NOS-II-derived NO in defense mechanisms cannot be ruled out and the discovery of an inducible NOS also in lower invertebrates will have implications for the emerging field of invertebrate ecological immunology.

Conclusions

The wide spectrum of physiological effects of NO in vertebrates prompted researchers to look for the presence of NO in invertebrates. NO was found to be a signaling molecule widespread throughout the metazoan kingdom and whose functions were highly conserved during evolution. These features were extended to the entire animal kingdom after the discovery of NOS activity in protozoa, yeasts and bacteria. Unfortunately, lack of detailed biochemical characterization of such a diverse protein family as invertebrate NOSs is one of the major limitations in the field. Future experiments using purified or expressed NOSs from various phyla are one of the crucial steps in our understanding of the evolution of NO signaling and its function. On the other hand, the great diversity of conventional NOSs within the animal kingdom clearly prevents interpretation of the emerging comparative data in terms of the existing classification of mammalian NOS isoforms. This classification might reflect the evolution of NOS isoforms within a defined lineage of vertebrates, but it needs to be readjusted as soon as novel cloning data from representatives of the more than 30 other animal phyla become available. The emerging molecular information about NOSs from lower invertebrates (porifera, cnidarians up to higher invertebrates) suggests that NO pathways might represent examples of parallel evolution of the NOS prototype in different lineages of animals. As a whole, all of the literature data indicate that NO is one of the earliest and most widespread signaling molecules in living organisms.

References

1. Moncada S, Palmer RM, Higgs EA. Nitric oxide: physiology, pathophysiology and pharmacology. Pharmacol Rev 1991; 43(2):109-142.
2. Garbers DL. Guanylyl cyclase receptors and their endocrine, paracrine and autocrine ligands. Cell 1992; 71(1):1-4.
3. Colasanti M, Suzuki H. The dual personality of NO. Trends Pharmacol Sci 2000; 21(7):249-252.
4. Colasanti M, Venturini G. Nitric oxide in invertebrates. Mol Neurobiol 1998; 17(1-3):157-174.
5. Elofsson R, Carlberg M, Moroz L et al. Is nitric oxide (NO) produced by invertebrate neurones? Neuroreport 1993; 4(3):279-282.
6. Jacklet JW. Nitric oxide signaling in invertebrates. Invert Neurosci 1997; 3(1):1-14.
7. Martinez A. Nitric oxide synthase in invertebrates. Histochem J 1995; 27(10):770-776.
8. Palumbo A. Nitric oxide in marine invertebrates: A comparative perspective. Comp Biochem Physiol A Mol Integr Physiol 2005; 142(2):241-248.
9. Johansson KU, Carlberg M. NO-synthase: what can research on invertebrates add to what is already known? Adv Neuroimmunol 1995; 5(4):431-442.
10. Moroz LL, Gillette R. From Polyplacophora to Cephalopoda: Comparative analysis of nitric oxide signalling in mollusca. Acta Biol Hung 1995; 46(2-4):169-182.
11. Giovine M, Pozzolini M, Favre A et al. Heat stress-activated, calcium-dependent nitric oxide synthase in sponges. Nitric Oxide 2001; 5(5):427-431.
12. Venturini G, Colasanti M, Fioravanti E et al. Direct effect of temperature on the catalytic activity of nitric oxide synthases types I, II and III. Nitric Oxide 1999; 3(5):375-382.
13. Zocchi E, Carpaneto A, Cerrano C et al. The temperature-signaling cascade in sponges involves a heat-gated cation channel, abscisic acid and cyclic ADP-ribose. Proc Natl Acad Sci USA 2001; 98(26):14859-14864.

14. Muller WE, Ushijima H, Batel R et al. Novel mechanism for the radiation-induced bystander effect: Nitric oxide and ethylene determine the response in sponge cells. Mutat Res 2006; 597(1-2):62-72.
15. Tran CT, Fox MF, Vallance P et al. Chromosomal localization, gene structure and expression pattern of DDAH1: Comparison with DDAH2 and implications for evolutionary origins. Genomics 2000; 68(1):101-105.
16. Ellwanger K, Nickel M. Neuroactive substances specifically modulate rhythmic body contractions in the nerveless metazoon tethya wilhelma (demospongiae, porifera). Front Zool 2006; 3:7.
17. Colasanti M, Lauro GM, Venturini G. NO in Hydra feeding response. Nature 1995; 374(6522):505.
18. Salleo A, Musci G, Barra P et al. The discharge mechanism of acontial nematocytes involves the release of nitric oxide. J Exp Biol 1996; 199(Pt 6):1261-1267.
19. Morrall CE, Galloway TS, Trapido-Rosenthal HG et al. Characterisation of nitric oxide synthase activity in the tropical sea anemone aiptasia pallida. Comp Biochem Physiol B Biochem Mol Biol 2000; 125(4):483-491.
20. Perez S, Weis V. Nitric oxide and cnidarian bleaching: An eviction notice mediates breakdown of a symbiosis. J Exp Biol 2006; 209(Pt 14):2804-2810.
21. Moroz LL, Meech RW, Sweedler JV et al. Nitric oxide regulates swimming in the jellyfish Aglantha digitale. J Comp Neurol 2004; 471(1):26-36.
22. Anctil M, Poulain I, pelletier C. Nitric oxide modulates peristaltic muscle activity associated with fluid circulation in the sea pansy Renilla koellikeri. J Exp Biol 2005; 208(Pt 10):2005-2017.
23. Colasanti M, Venturini G, Merante A et al. Nitric oxide involvement in Hydra vulgaris very primitive olfactory-like system. J Neurosci 1997; 17(1):493-499.
24. Moroz LL, Kohn AB. On the comparative biology of Nitric Oxide (NO) synthetic pathways: Parallel evolution of NO-mediated signaling. In: Tota B, Trimmer B, eds. Nitric Oxide. Amsterdam: Elsevier, 2007:1-44.
25. Breer H, Shepherd GM. Implications of the NO/cGMP system for olfaction. Trends Neurosci 1993; 16(1):5-9.
26. Eriksson KS. Nitric oxide synthase in the pharynx of the planarian Dugesia tigrina. Cell Tissue Res 1996; 286(3):407-410.
27. Gustafsson MK, Lindholm AM, Terenina NB et al. NO nerves in a tapeworm. NADPH-diaphorase histochemistry in adult Hymenolepis diminuta. Parasitology 1996; 113(Pt 6):559-565.
28. Terenina NB, Onufriev MV, Gulyaeva NV et al. A radiometric analysis of nitric oxide synthase activity in Hymenolepis diminuta. Parasitology 2000; 120(Pt 1):91-95.
29. Onufriev MV, Gulyaeva NV, Terenina NB et al. The effect of a nitric oxide donor on the synthesis of cGMP in Hymenolepis diminuta: a radiometric study. Parasitol Res 2005; 95(1):22-24.
30. Lindholm AM, Reuter M, Gustafsson MK. The NADPH-diaphorase staining reaction in relation to the aminergic and peptidergic nervous system and the musculature of adult Diphyllobothrium dentriticum. Parasitology 1998; 117(Pt 3):283-292.
31. Terenina NB, Reuter M, Gustafsson MK. An experimental, NADPH-diaphorase histochemical and immunocytochemical study of Mesocestoides vogae tetrathyridia. Int J Parasitol 1999; 29(5):787-793.
32. Gustafsson MK, Terenina NB, Kreshchenko ND et al. Comparative study of the spatial relationship between nicotinamide adenine dinucleotide phosphate-diaphorase activity, serotonin immunoreactivity and GYIRFamide immunoreactivity and the musculature of the adult liver fluke, Fasciola hepatica (digenea, fasciolidae). J Comp Neurol 2001; 429(1):71-79.
33. Terenina NB, Onufriev MV, Gulyaeva NV et al. Nitric oxide synthase activity in Fasciola hepatica: a radiometric study. Parasitology 2003; 126(Pt 6):585-590.
34. Kar PK, Tandon V, Saha N. Anthelmintic efficacy of flemingia vestita: genistein-induced effect on the activity of nitric oxide synthase and nitric oxide in the trematode parasite, fasciolopsis buski. Parasitol Int 2002; 51(3):249-257.
35. Tandon V, Kar PK, Saha N. NO nerves in trematodes, too! NADPH-diaphorase activity in adult fasciolopsis buski. Parasitol Int 2001; 50(3):157-163.
36. Terenina NB, Gustafsson MK. Nitric oxide and its target cells in cercaria of diplostomum chromatophorum: a histochemical and immunocytochemical study. Parasitol Res 2003; 89(3):199-206.
37. Kim SH, Chung JY, Bae YA et al. Functional identification of a protein inhibitor of neuronal nitric oxide synthase of taenia solium metacestode. Mol Biochem Parasitol 2007; 151(1):41-51.
38. Kohn AB, Moroz LL, Lea JM et al. Distribution of nitric oxide synthase immunoreactivity in the nervous system and peripheral tissues of schistosoma mansoni. Parasitology 2001; 122(Pt 1):87-92.
39. Long XC, Bahgat M, Chlichlia K et al. Detection of inducible nitric oxide synthase in schistosoma japonicum and S mansoni. J Helminthol 2004; 78(1):47-50.
40. Kohn AB, Lea JM, Moroz LL et al. Schistosoma mansoni: use of a fluorescent indicator to detect nitric oxide and related species in living parasites. Exp Parasitol 2006; 113(2):130-133.

41. Carolei A, Margotta V, Palladini G. Proposal of a new model with dopaminergic-cholinergic interactions for neuropharmacological investigations. Neuropsychobiology 1975; 1(6):355-364.
42. Venturini G, Stocchi F, Margotta V et al. A pharmacological study of dopaminergic receptors in planaria. Neuropharmacology 1989; 28(12):1377-1382.
43. Rawls SM, Rodriguez T, Baron DA et al. A nitric oxide synthase inhibitor (L-NAME) attenuates abstinence-induced withdrawal from both cocaine and a cannabinoid agonist (WIN 55212-2) in Planaria. Brain Res 2006; 1099(1):82-87.
44. Gustafsson MK, Terenina NB, Reuter M et al. NO nerves and their targets in a tapeworm: An immunocytochemical study of cGMP in hymenolepis diminuta. Parasitol Res 2003; 90(2):148-152.
45. Messerli SM, Morgan W, Birkeland SR et al. Nitric oxide-dependent changes in schistosoma mansoni gene expression. Mol Biochem Parasitol 2006; 150(2):367-370.

CHAPTER 7

Second Messenger Systems in Planaria

Scott M. Rawls*

Abstract

Second messengers are the cellular currency of information transfer. These signaling molecules, along with their related pathways, mediate the acute effects of drugs and the changes in neural plasticity caused by long-term drug exposure. Three of the best characterized second messengers are cyclic AMP (cAMP), nitric oxide (NO) and calcium ions. Although a plethora of functional roles for each of these signaling molecules have been unmasked in mammalian studies, the nature of the mechanisms, especially as related to intracellular signaling pathways within the central nervous system, remains poorly understood. An attractive alternative to mammals for investigating second messenger physiology is planarians. These flatworms possess a bilateral, yet simple, central nervous system; utilize mammalian-like neurotransmitters and signaling molecules; respond with characteristic behaviors to selective ligands; and display simpler pharmacokinetics related to the absorption, distribution, metabolism and excretion of drugs. This chapter reviews the documented functions of second messengers in mammals and humans, presents evidence of their functional impact on physiological and pathophysiological processes in planarians and speculates about how exploration of planarian second messenger pathways will contribute to a better understanding of human biology.

Second Messenger Overview

The concept of a second messenger system was established 50 years ago when Earl Sutherland and Ted Rall discovered that cyclic AMP (cAMP) mediates the intracellular actions of glucagon and epinephrine on glycogen metabolism in the liver. Calcium ions were the next signaling molecules to be identified as second messengers. The characterization of these second messengers has increased our knowledge about the importance of localized signaling events within a cell and the mechanism of communication between extracellular and intracellular signals. It is now widely accepted that second messengers play a critical role in signal transduction, a process by which a cell converts one kind of signal or stimulus into another type of signal. In this process, hormones and neurotransmitters are extracellular signals that activate a plasma membrane receptor, which functions as a link in communication between extracellular events and biochemical changes within a cell. The receptors signal their recognition of a bound ligand by initiating a series of reactions that culminates in a specific intracellular response. Second messenger molecules, such as calcium ions and cAMP and the original, extracellular messenger (i.e., hormone or neurotransmitter) are key components of a cascade of events that translates receptor activation by a hormone or neurotransmitter into a cellular response. Although cAMP and calcium ions were the first second messengers to be identified, a number of other second messengers, including nitric oxide and cyclic GMP (cGMP), have been identified.[1-3] Over the past five decades, the second messengers and their related pathways have been extensively characterized in mammalian models. Results

*Scott M. Rawls—Department of Pharmaceutical Sciences, Temple University School of Pharmacy, 3307 N. Broad Street, Philadelphia, Pennsylvania, USA.
Email: scott.rawls@temple.edu

Planaria: A Model for Drug Action and Abuse, edited by Robert B. Raffa and Scott M. Rawls.
©2008 Landes Bioscience.

from mammalian studies have identified a role for second messengers in nearly every physiological and pathophysiological process, including regeneration, feeding, intermediary metabolism, homeostasis, reproduction, digestion, learning, pain transmission and drug addiction. Although it is likely that planarians utilize mammalian-like second messengers, the precise role of these signaling molecules remains elusive. The current literature suggests that planarians utilize at least two second messengers—nitric oxide and cAMP.

Nitric Oxide

Nitric oxide (NO) is produced and released by many different types of cells in living organisms, primarily as a tool for intercellular communication.[4] Nitric oxide (NO) is recognized as a prominent second messenger.[5] production of this low-molecular weight, diffusible gas involves the enzyme nitric oxide synthase (NOS). The enzyme catalyzes the conversion of L-arginine into L-citrulline and NO. Prior work in mammals has shown that NOS is found in peripheral and central neurons.[6] Three different isoforms of NOS have been described.[7] Two are constitutive forms, endothelial and neuronal and the third is inducible.[8] Neuronal NOS (nNOS) was first described in the neurons of the central and peripheral nervous systems and eNOS is generally found in the endothelium of blood vessels responsible for vasodilatation. The third isoform, iNOS, is located in the cytosol of cells in the immune system, in smooth muscles and hepatocytes. It was originally described as inducible and is almost undetectable under basal conditions but is induced at the transcriptional level by cytokines.[9] NO plays an important modulator role not only in the cardiovascular system but also in other physiological and pathophysiological processes, including thermoregulation, fever, hyperthermia, learning, aggression and drug addiction.[10-12] With regard to drug addiction, most of the studies in mammals have focused on a role for nitric oxide production in opioid dependence, tolerance and reward. It has been documented that the inhibition of nitric oxide production prevents the development, expression and maintenance of physical dependence in mammals.[13-19] These studies have led to the belief that nitric oxide production is increased during the development of opioid tolerance and dependence andw that nitric oxide production is required for the maintenance of a preexisting opioid dependence. Nitric oxide synthase inhibitors also block morphine, cocaine, nicotine and methamphetamine conditioned place preference, an assay commonly used to evaluate the rewarding effects of drugs in mammals.[20-22]

The physiological actions of nitric oxide in the mammalian brain are linked closely to glutamate and calcium ions. Glutamate is a major excitatory neurotransmitter and it has a well-documented role in opioid dependence and tolerance and in the rewarding and reinforcing effects of pyschostimulants such as cocaine, methamphetamine and amphetamine.[14,15,23-27] The activation of NMDA receptors by glutamate triggers the influx of calcium ions, which, in turn, bind to calmodulin and cause the activation of nNOS.[6,28,29] This mode of activation explains the ability of glutamatergic neurotransmission to stimulate NO formation in a matter of seconds and underscores the fact that the production of NO is a calmodulin-dependent process that must be preceded by the elevation of intracellular calcium ion concentration. As a wide range of receptors might mediate effects that result in an elevation in the intracellular calcium ion concentration, the close relationship between NMDA receptors and NO synthesis seems puzzling. Clarification of the molecular organization of glutamatergic synapses provides an explanation for this phenomenon. Neuronal NOS is connected to the NMDA receptors via a postsynaptic density protein, thus ensuring that the enzyme is directly exposed to the flux of calcium ions entering the ion channel of activated NMDA receptors.[29] Calcium ion transients that arise from the activation of other receptors are presumably too diluted by the time they reach the vicinity of the enzyme. nNOS can therefore only be 'switched on' by NMDA receptors. Consequently, the level of endogenously produced NO around the NMDA receptor-nNOS-containing synapses reflects the activity of glutamate-mediated neurotransmission.

In addition to its critical role in controlling synaptic activity, NO is also an ideal mediator of extrasynaptic interactions because of its physicochemical properties.[30,31] It is a highly diffusible gas that easily penetrates biological membranes. Although its half-life is only a few seconds, even during this short period it can diffuse a few hundred micrometers.[32] Comparing this distance with the width of a synaptic cleft (20 nm) or the size of a cell body (a few micrometers), it is evident that NO produced postsynaptically by nNOS might influence the function of a large number of neurons in a sphere around the synapse.

Nitric Oxide and Planarians

The precise second messengers which play a role in planarian physiology have not been identified, but evidence suggests that NO is a likely candidate. For example, NO is unique among signaling molecules in the sense that it is nonpolar enough to freely cross the plasma membranes of cells and soluble enough in aqueous media to diffuse through the intracellular and extracellular compartments. Clearly, these physiobiochemical properties (e.g., diffusible, gaseous, low-molecular weight, nonpolar) are well suited for the aqueous habitat of planarians. A more direct role for nitric oxide production in planarian physiology was provided in elegant experiments by Eriksson and colleagues.[25] In that study, the distribution of the nicotinamide adenine dinucleotide phosphate-(NADPH) diaphorase reaction, an indicator of nitric oxide synthase activity, was studied in the pharynx of the freshwater planarian Dugesia tigrina (Platyhelminthes). High pressure liquid chromatography (HPLC) was used to measure the formed citrulline, which is an indirect measure of NOS activity. The activity of the enzyme was found to be dependent on NADPH, whereas no dependency on calcium ions and calmodulin was detected. Because enzyme activity was measured in the pharynx, Eriksson concluded that NO and NOS has an important role in the feeding behavior of planarians. One important difference between NOS activity in planarians and mammals appears to be the degree of calcium ion-sensitivity. It is well documented that nNOS in mammals is highly dependent on an increase in the intracellular calcium ion concentration and subsequent formation of a calcium-calmodulin complex.[28,29] observation that NOS activity in planarians is insensitive to calcium ions and calmodulin, suggests that planarians utilize a form of NOS that is activated by signals that are not entirely the same as those signals which trigger NO production in mammals.[25] NO production is required for the expression of cannabinoid and cocaine physical dependence in planarians.[33] A description of the experiments and results leading to this conclusion is presented below.

Planarians (Dugesia dorotocephala) were purchased from Carolina Biological Supply Co. (Burlington, NC), acclimated to temperature-controlled room temperature (21°C) and tested within 72 h. Each planarian was used only once. Cocaine, WIN 55212-2 ((R)-(+)-[2,3-Dihydro-5 methyl-3-[(4 morpholino)methyl]pyrrolo-[1,2,3-de]-1,4 benzoxazin-6-yl](1-naphthyl)metha-none), WIN 55212-3, L-NAME (l-nitro-arginine methyl ester) and cremophor were purchased from Sigma Chemical Co. (St. Louis, MO). To quantify planarian locomotor velocity (pLMV),[34] individual planarians were placed into a clear plastic petri dish (14 cm diameter) containing room-temperature (21°C) tap water treated with AmQuel® water conditioner. The dish was located over paper with gridlines spaced 0.5 cm apart. pLMV was quantified as the number of gridlines planarians crossed or recrossed per minute over a 5 min observation period and is expressed as the mean (± SD) of the cumulative number of gridlines crossed by each planarian per minute. The metric used to quantify abstinence-induced withdrawal is a change in planarian locomotor velocity (pLMV), which is described in detail in Chapter 8. Prior to measurement of pLMV, each planarian was placed into individual 0.5 ml vials containing room-temperature vehicle or test compound(s) for 1 h. For WIN 552122-2 and WIN 55212-3, a stock solution of 1 mM was prepared in 10/90% cremophor/water and treatment solutions diluted with water. Each planarian was exposed individually for 1 h to one of the following treatments: water, cocaine (80 μM), WIN 55212-2 (10 μM), WIN 55212-3 (10 μM), L-NAME (10 μM) and then tested individually for pLMV over 5 min in one of the following: water, cocaine (80 μM), WIN 55212-2 (10 μM), WIN 55212-3 (10 μM), or L-NAME (10 μM).

L-NAME (10 μM) had no effect of its own on pLMV, whether the planarians were exposed to it during the pretreatment phase (60 min), the test phase (5 min), or both (Fig. 1). However, planarians pretreated with WIN 55212-2 (10 μM) displayed significantly greater pLMV (ANOVA: P

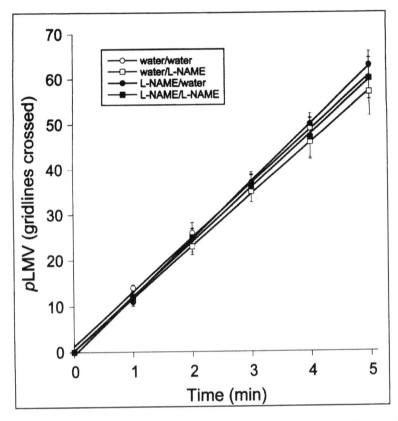

Figure 1. Lack of effect of a NOS inhibitor, L-NAME, on planarian locomotor velocity (pLMV), expressed as the cumulative number of gridline crossings, of planarians pretreated in water (water/) or 10 μM L-NAME (L-NAME/) and tested in water (/water) or 10 μM L-NAME (/L-NAME). Reprinted with permission from reference 33.

< 0.0001, F = 71.03) when tested in 10 μM L-NAME than when tested in water (Fig. 2). Hence, L-NAME significantly attenuated the abstinence-induced withdrawal from WIN 55212-2 in this model. The results with cocaine are shown in Fig. 3. Planarians exposed to cocaine for 1 h and then placed into cocaine-free water displayed a significantly reduced pLMV. As was the case with WIN 55212-2, the pLMV remained constant over the 5-min observation period, even during withdrawal from cocaine, suggesting a physiologic, not toxic or local anesthetic, effect of cocaine on the planarians.

The presence of a withdrawal syndrome following the cessation of drug exposure remains the hallmark indicator of physical dependence in a living organism. On the basis of this definition, our data indicate that nitric oxide production is required for the expression of both cannabinoid and cocaine physical dependence in planarians.[33] In the presence of a drug that inhibits nitric oxide production, such as L-NAME, the expression of abstinence-induced withdrawal is inhibited.[33] Two important considerations of these results are the consistency of the data with mammalian models and the exact mechanism. Experiments in rodent models have established that NO pathways contribute to neuronal adaptations in response to repeated exposure to cocaine, morphine, or alcohol and that the inhibition of NO production attenuates morphine withdrawal symptoms and cocaine sensitization in rats and mice[14-16,19,35-38] and guinea pig isolated ileum preparations.[39] Furthermore, central nitric oxide synthase (NOS) expression increases following withdrawal from repeated exposure to morphine or cocaine in rats.[36,40-42] Those collective results suggest that

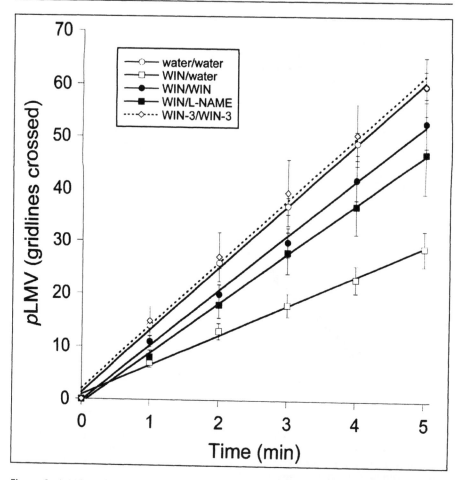

Figure 2. A NOS inhibitor, L-NAME, inhibits abstinence-induced withdrawal caused by a cannabinoid agonist, WIN 55212-2. Planarian locomotor velocity (pLMV), expressed as the cumulative number of gridline crossings, of planarians pretreated in water (water/) or 10 μM WIN 55212-2 (WIN/) and tested in water (/water), 10 μM WIN 55212-2 (/WIN) or 10 μM L-NAME (/L-NAME). Dotted line is 10 μM WIN 55212-3 (WIN-3), the inactive enantiomer of WIN 55212-2. Reprinted with permission from reference 33.

inhibiting NO production may be a common mechanism to attenuate withdrawal from a variety of abused substances. Because withdrawal behaviors following cessation of exposure to cannabinoids and cocaine are difficult to quantify in mammals, our knowledge about a role for nitric oxide in the specific processes of cannabinoid and cocaine physical dependence in mammals is limited. Planarians might represent a more sensitive model for development of physical dependence to these agents.

The exact process by which NO production promotes physical dependence in planarians is not known at the current time, but an interaction with the glutamate system is likely to play a role. The intimate association between glutamate and NO in the regulation of numerous physiological and pathophysiological processes, in addition to their involvement in physical dependence and tolerance, was discussed in Section 6.2. It is now becoming clear that planarians, as well as mammals, utilize glutamate and its associated NMDA-like receptors. HPLC analysis has shown that planarians contain both glutamate and aspartate.[43] Molecular studies have revealed that planarians

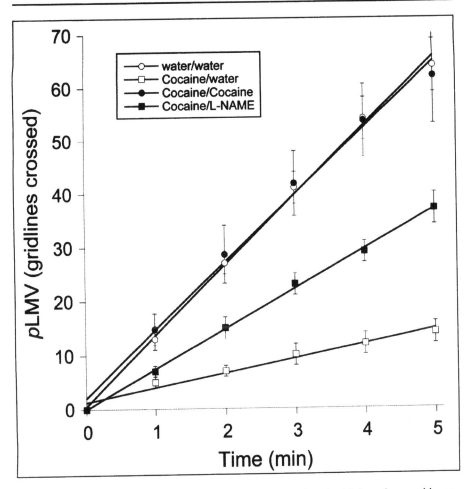

Figure 3. A NOS inhibitor, L-NAME, inhibits abstinence-induced withdrawal caused by cocaine. Planarian locomotor velocity ($pLMV$), expressed as the cumulative number of gridline crossings, of planarians pretreated in water (water/) or 80 μM cocaine (Cocaine/) and tested in water (/water), 80 μM cocaine (/Cocaine) or 10 μM L-NAME (/L-NAME). Reprinted with permission from reference 33.

express the genes for at least two types of ionotropic glutamate receptors which share high sequence similarity to neural specific genes isolated from humans and mice.[44] Similar to a NOS inhibitor, a NMDA antagonist (LY 235959) blocks the development (Fig. 4) and expression (Fig. 5) of cannabinoid physical dependence in planarians.[45] This result suggests that the activation of NMDA-like receptors in planarians is required for cannabinoids to produce physical dependence. Given the close association between NMDA receptor activation and NO production in mammals, it is conceivable that the concurrent activation of NMDA and NO systems, at least in planarians, is a common mechanism through which a broad array of drugs produces physical dependence.

cAMP and CREB

CREB (cAMP response element binding protein) is a member of the bZIP superfamily of transcription factors. It is composed of a C-terminal basic domain that is responsible for binding to DNA and a leucine zipper domain that mediates dimerzation with itself or other members of the CREB family of transcription factors, including CREM (cAMP response element modulator)

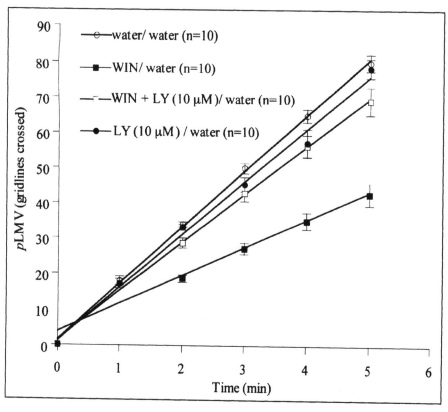

Figure 4. A NMDA receptor antagonist, LY 235959 (LY), blocks the development of physical dependence to a cannabinoid agonist, WIN 55212-2 (WIN). Planarians were pretreated in one of the following: water; WIN (10 μM); LY (10 μM); or a combination of WIN (10 μM) and LY (10 μM) for 60 min. Planarians were then tested in water for 5 min. Planarian locomotor velocity (pLMV) was quantified as the number of gridlines crossed or recrossed per minute over a 5-minute interval and expressed as the mean ± S.E.M. of the cumulative number of gridlines crossed by each planarian per minute. Reprinted with permission from reference 45.

and ATF-1 (activating transcription factor 1). CREB dimers bind to the CRE (consensus cAMP response element). Many genes have CRE sites in their promoters, including neuropeptides, neurotransmitter synthesizing enzymes, neurotransmitter receptors, signaling proteins and other transcription factors.[46,47] Phosphorylation and subsequent activation of CREB is a site of convergence for several signal transduction cascades, including the cAMP pathway via protein kinase A (PKA), intracellular calcium ions via calcium calmodulin-dependent kinases (CaMKII), the Ras/extracellular signal regulated kinase (ERK) protein kinase pathway, the phosphotidylinositol-3-kinase (PI3K) kinase pathway and stress-induced signaling cascades.[47] CREB binding protein (CBP) subsequently binds to the phosphorylated CREB dimer and serves as an adaptor to the transcription initiation complex. The histone acetyltransferase (HAT) activity endogenous to CBP unravels chromatin and facilitates transcription.[48]

Mammalian studies have revealed that CREB plays a key role in drug addition. The activation of CREB is downstream of the cAMP signaling pathway, a pathway that is upregulated as an adaptation to chronic exposure to drugs of abuse.[48] Chronic exposure to opioids has been shown to upregulate the cAMP signaling cascade.[49-53] This upregulation is thought to be a compensatory response to the acute inhibitory actions of opiates, which bind to G_i-coupled receptors and inhibit cAMP production inside the cell. Upregulation of the cAMP pathway mediates several aspects of

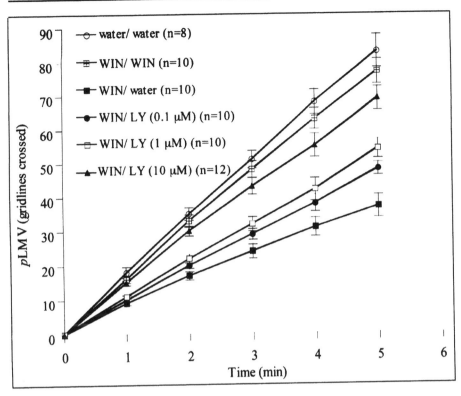

Figure 5. A NMDA receptor antagonist, LY 235959 (LY), blocks the expression of physical dependence to a cannabinoid agonist, WIN 55212-2 (WIN). Planarians were pretreated in water or WIN (10 µM) for 60 min and then tested in one of the following: water; WIN (10 µM); or LY (0.1, 1 or 10 µM) for 5 min. Planarian locomotor velocity (pLMV) was quantified as the number of gridlines crossed or recrossed per minute over a 5-minute interval and expressed as the mean ± S.E.M. of the cumulative number of gridlines crossed by each planarian per minute. Reprinted with permission from reference 45.

addiction, depending on the specific brain region involved.[54] The role of the cAMP pathway in the locus coeruleus (LC) and nucleus accumbens (NAC) has been extensively characterized. Cyclic AMP signaling within the LC plays an important role in the compensatory responses to repeated opioid exposure. The LC is located at the base of the fourth ventricle. It is the major noradrenergic nucleus in the brain and its primary functions are to control the sympathetic nervous system and attention.[55-57] The upregulation of the cAMP pathway and CREB, in the LC underlies some of the symptoms which of opioid physical dependence and withdrawal.[49,58] The cAMP signaling pathway is inhibited following acute opioid exposure; however, the signaling cascade, including the expression of CREB, is increased in the LC following repeated morphine exposure.[59] Among the genes involved are adenylyl cyclase type VIII and tyrosine hydroxylase, whose expression is upregulated by chronic morphine administration via a CREB-dependent mechanism.[60] Mice containing mutations of the CREB gene display reduced symptoms of morphine withdrawal and a strong aversion to opioid withdrawal in a conditioned-aversion paradigm despite their attenuated physical withdrawal symptoms.[61,62] This suggests that cAMP-related mechanisms of physical dependence and negative motivational aspects of morphine withdrawal are not entirely the same.[59]

  The cAMP signaling pathway is also upregulated in the NAC following repeated exposure to opioids, cocaine and alcohol.[49,63,64] Activation of CREB and CRE-mediated transcription has also been observed in response to chronic morphine and amphetamine treatments in the NAC.[52,53,65]

Beyond attenuating the rewarding effects of drugs of abuse, upregulation of the cAMP pathway and CREB in the NAC may also contribute to states of dysphoria seen early in withdrawal.[66] Thus, CREB overexpression in the NAC, achieved with viral vectors or in inducible transgenic mice, produces depression-like responses in the forced-swim and learned-helplessness tests, whereas mutant CREB expression causes antidepressant-like responses.[67] Efforts are currently under way to identify target genes for CREB in the NAC. One apparent target is dynorphin, an opioid neuropeptide expressed in a subset of medium spiny neurons in the NAC. Neurochemical evidence reveals that dynorphin binds to kappa opioid receptors on dopamine neuronal cell bodies and terminals, in the midbrain ventral tegmental area to inhibit their activity and decrease the release of dopamine from their terminals in the NAC.[68] Dynorphin in the NAC is induced following chronic drug exposure,[69,70] and a kappa opioid antagonist decreases both the aversive and depression-like responses to cocaine are decreased with a kappa opioid antagonist.[69]

cAMP in Planarians

Several lines of evidence suggest that planarians utilize cAMP for a number of physiological processes. For example, a prior study demonstrated that cAMP inhibits regeneration in planarians whereas a related cyclic nucleotide, cGMP, facilitates regeneration.[71,72] A role for cAMP in dopamine-mediated responses in planarians has also been reported.[73] Dopamine D_1 selective agonists produce screw-like hyperkinesias in planarians and this behavior is prevented by a D_1 receptor antagonist (SCH 23390) but not by a D_2 antagonist (sulpiride). On the other hand, dopamine D_2 receptor selective agonists produce a 'C-like' curling that is inhibited by pretreatment with D_2 selective antagonists. Dopamine D_1-selective agonists, mixed action agonists or D_2-selective agonists all induce a significant increase in cAMP levels that is prevented by pretreatment with a specific dopamine antagonist.[73] These data suggest that cAMP is present in planarians and that an increase in cAMP levels downstream of dopamine receptor activation contributes to dopamine-related behaviors. The localization of adenylate-cyclase, the enzyme which catalyzes the conversion of ATP into cAMP, has been studied in planarians. Molecular studies have revealed that cAMP is localized to specific regions of planarians, including mucous gland cells, in rhabdite cells, in intercellular spaces and in nerve endings.[38] It is possible that the cAMP present in nerve endings is responsible for regulating dopamine behaviors, particularly since dopamine released into the extracellular compartment of the mammalian brain arises exclusively from nerve terminals.

Our laboratory has been investigating a role of cAMP in physical dependence and withdrawal in planarians. Evidence indicates that an upregulation of the cAMP signaling cascade contributes to opioid physical dependence and withdrawal in mammals (See 6.4). However, the role of cAMP in the dependence caused by other drugs of abuse remains unclear, primarily because it is difficult to demonstrate quantifiable behaviors following the spontaneous withdrawal of mammals from many non-opioid drugs (e.g., methamphetamine, amphetamine, cocaine, etc.). In our studies, we initially investigated the effect of a cAMP inhibitor, cAMPS-Rp, triethylammonium salt, on abstinence-induced withdrawal from methamphetamine (Fig. 6). Experiments revealed that methamphetamine (10 μM), similar to other addictive drugs, produced an abstinence-related withdrawal syndrome.[74] This was apparent because methamphetamine-exposed planarians tested in water displayed locomotor activity that was significantly less than methamphetamine-naïve planarians tested in water (Fig. 6). The reduction in locomotor activity (i.e., withdrawal) was almost completely abolished when methamphetamine-exposed were tested in a cAMP antagonist (Fig. 6). These data suggest that the expression of methamphetamine physical dependence in planarians requires an increase in cAMP. Current studies are underway to determine whether a role for increased cAMP extends to the physical dependence caused by other drugs of abuse.

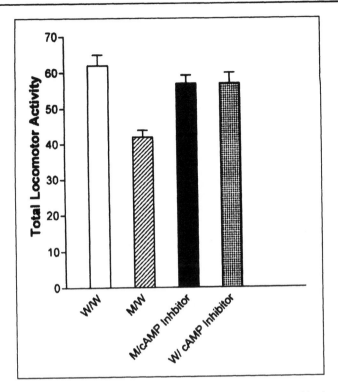

Figure 6. A cyclic AMP (cAMP) inhibitor (cAMPS-Rp-triethylammonium) blocks the expression of methamphetamine physical dependence. Planarians were pre-treated in water (w) or methamphetamine (M) (10 µM) for 60 min and then tested in either water or the cAMP inhibitor (10 µM) for 5 min. Total locomotor activity was quantified as the number of gridlines crossed or re-crossed per minute over a 5-minute interval and expressed as the mean ± S.E.M. of the cumulative number of gridlines crossed in 5 minutes.

Conclusions

Given the degree of conservation between planarian and mammalian genes, the investigation of second messengers in planarians should contribute to a better understanding of the functional role these signaling molecules play in human physiology and pathophysiology. It is already known that at least two mammalian-like second messengers, nitric oxide and cAMP, contribute to hyperactivity, regeneration, physical dependence and withdrawal in planarians and future research seems destined to identify a role for additional signaling molecules in planarian biology. The sensitive behavioral responses of planarians to drugs and their utilization of mammalian-like neurotransmitters, combined with modern methodologies for studying intracellular signaling, indicate that planarians will be an excellent system for studying the physiological roles of second messengers and their associated signaling pathways.

References

1. Feelisch M, te Poel M, Zamora R et al. Understanding the controversy over the identity of EDRF. Nature 1994; 368:62-65.
2. Griffith TM, Edwards DH, Lewis MJ et al. The nature of endothelium-derived vascular relaxant factor. Nature 1984; 308:645-647.
3. Palmer RM, Ferrige AG, Moncada S. Nitric oxide release accounts for the biological activity of endothelium-derived relaxing factor. Nature 1987; 327:524-526.
4. Lancaster JR Jr. A tutorial on the diffusibility and reactivity of free nitric oxide. Nitric Oxide 1997; 1:18-30.

5. Breder CD, Saper CB. Expression of inducible cyclooxygenase mRNA in the mouse brain after systemic administration of bacterial lipopolysaccharide. Brain Res 1996; 713:64-69.
6. Bredt DS, Snyder SH. Nitric oxide, a novel neuronal messenger. Neuron 1992; 8(1):3-11.
7. López-Figueroa MO, Itoi K, Watson SJ. Regulation of nitric oxide synthase messenger RNA expression in the rat hippocampus by glucocorticoids. Neuroscience 1998; 87:439-446.
8. Lowenstein CJ, Glatt CS, Bredt DS et al. Cloned and expressed macrophage nitric oxide synthase contrasts with the brain enzyme. Proc Natl Acad Sci USA 1992; 89:6711-6715.
9. Xie QW, Cho HJ, Calaycay J et al. Cloning and characterization of inducible nitric oxide synthase from mouse macrophages. Science 1992; 256:225-228.
10. Amir S, De Blasio E, English AM. NG-monomethyl-L-arginine co-injection attenuates the thermogenic and hyperthermic effects of E2 prostaglandin microinjection into the anterior hypothalamic preoptic area in rats. Brain Res 1991; 556:157-160.
11. Gerstberger R. Nitric Oxide and Body Temperature Control. News Physiol Sci 1999; 14:30-36.
12. Steiner DR, Gonzalez NC, Wood JG. Interaction between reactive oxygen species and nitric oxide in the microvascular response to systemic hypoxia. J Appl Physiol 2002; 93:1411-1418.
13. Bhargava HN. Attenuation of tolerance to and physical dependence on, morphine in the rat by inhibition of nitric oxide synthase. Gen Pharmacol 1995; 26:1049-1053.
14. Cappendijk SL, de Vries R, Dzoljic MR. Excitatory amino acid receptor antagonists and naloxone-precipitated withdrawal syndrome in morphine-dependent mice. Eur Neuropsychopharmacol 1993; 3:111-116.
15. Cappendijk SL, de Vries R, Dzoljic MR. Inhibitory effect of nitric oxide (NO) synthase inhibitors on naloxone-precipitated withdrawal syndrome in morphine-dependent mice. Neurosci Lett 1993; 162:97-100.
16. Kimes AS, Vaupel DB, London ED. Attenuation of some signs of opioid withdrawal by inhibitors of nitric oxide synthase. Psychopharmacology 1993; 112:521-524.
17. Kolesnikov YA, Pick CG, Ciszewska G et al. Blockade of tolerance to morphine but not to kappa opioids by a nitric oxide synthase inhibitor. Proc Natl Acad Sci USA 1993; 90:5162-5166.
18. Thorat SN, Bhargava HN. Effects of NMDA receptor blockade and nitric oxide synthase inhibition on the acute and chronic actions of delta 9-tetrahydrocannabinol in mice. Brain Res 1994; 667:77-82.
19. Vaupel DB, Kimes AS, London ED. Nitric oxide synthase inhibitors. Preclinical studies of potential use for treatment of opioid withdrawal. Neuropsychopharmacology 1995; 13:315-322.
20. Kivastik T, Rutkauskaite J, Zharkovsky A. Nitric oxide synthesis inhibition attenuates morphine-induced place preference. Pharmacol Biochem Behav 1996; 53:1013-1015.
21. Li SM, Yin LL, Shi J et al. The effect of 7-nitroindazole on the acquisition and expression of D-methamphetamine-induced place preference in rats. Eur J Pharmacol 2002; 435:217-223.
22. Martin JL, Itzhak Y. 7-Nitroindazole blocks nicotine-induced conditioned place preference but not LiCl-induced conditioned place aversion. Neuroreport 2000; 11:947-949.
23. Ben-Eliyahu S, Marek P, Vaccarino AL et al. The NMDA receptor antagonist MK-801 prevents long-lasting non-associative morphine tolerance in the rat. Brain Res 1992; 575:304-308.
24. Danbolt NC. Glutamate uptake. Prog Neurobiol 2001; 65:1-105.
25. Eriksson KS. Nitric oxide synthase in the pharynx of the planarian Dugesia tigrina. Cell Tissue Res 1996; 286:407-410.
26. Garthwaite J, Charles SL, Chess-Williams R. Endothelium-derived relaxing factor release on activation of NMDA receptors suggests role as intercellular messenger in the brain. Nature 1988; 336:385-388.
27. Rasmussen K, Krystal JH, Aghajanian GK. Excitatory amino acids and morphine withdrawal: differential effects of central and peripheral kynurenic acid administration. Psychopharmacology 1991; 105:508-512.
28. Bredt DS, Snyder SH. Nitric oxide: a physiologic messenger molecule. Annu Rev Biochem 1994; 63:175-195.
29. Brenman JE, Bredt DS. Synaptic signaling by nitric oxide. Curr Opin Neurobiol 1997; 7:374-378.
30. Schulman H. Nitric oxide: a spatial second messenger. Mol Psychiatry 1997; 2:296-299.
31. Vizi ES. Role of high-affinity receptors and membrane transporters in nonsynaptic communication and drug action in the central nervous system. Pharmacol Rev 2000; 52:63-89.
32. Gally JA, Montague PR, Reeke GN Jr et al. The NO hypothesis: possible effects of a short-lived, rapidly diffusible signal in the development and function of the nervous system. Proc Natl Acad Sci USA 1990; 87:3547-3551.
33. Rawls SM, Rodriguez T, Baron DA et al. A nitric oxide synthase inhibitor (L-NAME) attenuates abstinence-induced withdrawal from both cocaine and a cannabinoid agonist (WIN 55212-2) in Planaria. Brain Res 2006; 1099:82-87.

34. Raffa RB, Holland LJ, Schulingkamp RJ. Quantitative assessment of dopamine D2 antagonist activity using invertebrate (Planaria) locomotion as a functional endpoint. J Pharmacol Toxicol Methods 2001; 45:223-226.
35. Adams ML, Kalicki JM, Meyer ER et al. Inhibition of the morphine withdrawal syndrome by a nitric oxide synthase inhibitor, NG-nitro-L-arginine methyl ester. Life Sci 1993; 52:PL245-249.
36. Bhargava HN, Kumar S. Sensitization to the locomotor stimulant activity of cocaine is associated with increases in nitric oxide synthase activity in brain regions and spinal cord of mice. Pharmacology 1997; 55:292-298.
37. Byrnes JJ, Pantke MM, Onton JA et al. Inhibition of nitric oxide synthase in the ventral tegmental area attenuates cocaine sensitization in rats. Prog Neuropsychopharmacol Biol Psychiatry 2000; 24:261-273.
38. Panikkar GP. Cocaine Addiction: Neurobiology and Related Current Research in Pharmacotherapy. Subst Abus 1999; 20:149-166.
39. Capasso A, Sorrentino L, Pinto A. The role of nitric oxide in the development of opioid withdrawal induced by naloxone after acute treatment with mu- and kappa-opioid receptor agonists. Eur J Pharmacol 1998; 359:127-131.
40. Cuéllar B, Fernández AP, Lizasoain I et al. Up-regulation of neuronal NO synthase immunoreactivity in opiate dependence and withdrawal. Psychopharmacology 2000; 148:66-73.
41. Liang DY, Clark JD. Modulation of the NO/CO-cGMP signaling cascade during chronic morphine exposure in mice. Neurosci Lett 2004; 365:73-77.
42. Wong CS, Hsu MM, Chou YY et al. Morphine tolerance increases [3H]MK-801 binding affinity and constitutive neuronal nitric oxide synthase expression in rat spinal cord. Br J Anaesth 2000; 85:587-591.
43. Rawls SM, Gomez T, Stagliano GW et al. Measurement of glutamate and aspartate in Planaria. J Pharmacol Toxicol Methods 2006; 53:291-295.
44. Cebrià F, Kudome T, Nakazawa M et al. The expression of neural-specific genes reveals the structural and molecular complexity of the planarian central nervous system. Mech Dev 2002; 116:199-204.
45. Rawls SM, Gomez T, Raffa RB. An NMDA antagonist (LY 235959) attenuates abstinence-induced withdrawal of planarians following acute exposure to a cannabinoid agonist (WIN 55212-2). Pharmacol Biochem Behav 2007; 86:499-504.
46. Lonze BE, Ginty DD. Function and regulation of CREB family transcription factors in the nervous system. Neuron 2002; 35:605-623.
47. Mayr B, Montminy M. Transcriptional regulation by the phosphorylation-dependent factor CREB. Nat Rev Mol Cell Biol 2001; 2:599-609.
48. Guitart X, Thompson MA, Mirante CK et al. Regulation of cyclic AMP response element-binding protein (CREB) phosphorylation by acute and chronic morphine in the rat locus coeruleus. J Neurochem 1992; 58:1168-1171.
49. Nestler EJ, Aghajanian GK. Molecular and cellular basis of addiction. Science 1997; 278:58-63.
50. Sharma SK, Klee WA, Nirenberg M. Dual regulation of adenylate cyclase accounts for narcotic dependence and tolerance. Proc Natl Acad Sci USA 1975; 72:3092-3096.
51. Sharma SK, Nirenberg M, Klee WA. Morphine receptors as regulators of adenylate cyclase activity. Proc Natl Acad Sci USA 1975; 72:590-594.
52. Shaw-Lutchman TZ, Barrot M, Wallace T et al. Regional and cellular mapping of cAMP response element-mediated transcription during naltrexone-precipitated morphine withdrawal. J Neurosci 2002; 22:3663-3672.
53. Shaw-Lutchman TZ, Impey S, Storm D et al. Regulation of CRE-mediated transcription in mouse brain by amphetamine. Synapse 2003; 48:10-17.
54. Nestler EJ. Molecular neurobiology of addiction. Am J Addict 2001; 10:201-217.
55. Aston-Jones G. Brain structures and receptors involved in alertness. Sleep Med 2005;(Suppl 1):S3-7.
56. Foote SL, Bloom FE, Aston-Jones G. Nucleus locus ceruleus: new evidence of anatomical and physiological specificity. Physiol Rev 1983; 63:844-914.
57. Van Bockstaele EJ, Aston-Jones G. Integration in the ventral medulla and coordination of sympathetic, pain and arousal functions. Clin Exp Hypertens 1995; 17:153-165.
58. Lane-Ladd SB, Pineda J, Boundy VA et al. CREB (cAMP response element-binding protein) in the locus coeruleus: biochemical, physiological and behavioral evidence for a role in opiate dependence. J Neurosci 1997; 17:7890-7901.
59. Widnell KL, Russell DS, Nestler EJ. Regulation of expression of cAMP response element-binding protein in the locus coeruleus in vivo and in a locus coeruleus-like cell line in vitro. Proc Natl Acad Sci USA 1994; 91:10947-10951.
60. Chao JR, Ni YG, Bolaños CA et al. Characterization of the mouse adenylyl cyclase type VIII gene promoter: regulation by cAMP and CREB. Eur J Neurosci 2002; 16:1284-1294.
61. Maldonado R, Blendy JA, Tzavara E et al. Reduction of morphine abstinence in mice with a mutation in the gene encoding CREB. Science 1996; 273:657-659.

62. Maldonado R, Valverde O, Garbay C et al. Protein kinases in the locus coeruleus and periaqueductal gray matter are involved in the expression of opiate withdrawal. Naunyn Schmiedebergs Arch Pharmacol 1995; 352:565-575.

63. Terwilliger RZ, Beitner-Johnson D, Sevarino KA et al. A general role for adaptations in G-proteins and the cyclic AMP system in mediating the chronic actions of morphine and cocaine on neuronal function. Brain Res 1991; 548:100-110.

64. Unterwald EM, Cox BM, Kreek MJ et al. Chronic repeated cocaine administration alters basal and opioid-regulated adenylyl cyclase activity. Synapse 1993; 15:33-38.

65. Barrot M, Olivier JD, Perrotti LI et al. CREB activity in the nucleus accumbens shell controls gating of behavioral responses to emotional stimuli. Proc Natl Acad Sci USA 2002; 99:11435-11440.

66. Hyman SE, Malenka RC. Addiction and the brain: the neurobiology of compulsion and its persistence. Nat Rev Neurosci 2001; 2:695-703.

67. Pliakas AM, Carlson RR, Neve RL et al. Altered responsiveness to cocaine and increased immobility in the forced swim test associated with elevated cAMP response element-binding protein expression in nucleus accumbens. J Neurosci 2001; 21:7397-7403.

68. Spanagel R, Herz A, Shippenberg TS. Opposing tonically active endogenous opioid systems modulate the mesolimbic dopaminergic pathway. Proc Natl Acad Sci USA 1992; 89:2046-2050.

69. Carlezon WA Jr, Thome J, Olson VG et al. Regulation of cocaine reward by CREB. Science 1998; 282:2272-2275.

70. Daunais JB, McGinty JF. Acute and chronic cocaine administration differentially alters striatal opioid and nuclear transcription factor mRNAs. Synapse. 1994; 18(1):35-45.

71. Duma A. Activity in cAMP phosphodiesterase in the early regeneration stage of planarian Dugesia lugubris (O. Schmidt). Ultracytochemical studies. Acta Med Pol 1980; 21:317-318.

72. Lenicque PM. Control of the regeneration of small pieces of the planarian Dugesia tigrina by cylic AMP and GMP nucleotides. CR Acad Sci Hebd Seances Acad Sci D 1976; 283(11 D):1317-1319.

73. Venturini G, Stocchi F, Margotta V et al. Pharmacological study of dopaminergic receptors in planaria. Neuropharmacology 1989; 28:1377-1382.

74. Raffa RB, Stagliano GW, Ross G et al. The kappa-opioid receptor antagonist nor-BNI inhibits cocaine and amphetamine, but not cannabinoid (WIN 52212-2), abstinence-induced withdrawal in planarians: An instance of 'pharmacologic congruence'. Brain Res 2008; 1193:51-56

CHAPTER 8

Planaria as Model in Drug Abuse Research

Robert B. Raffa*

Abstract

In some complex way, drug abuse is a physiological process that involves the modulation or alteration of one or more neurochemical pathways. These pathways are reviewed in this chapter to the extent that involvement of particular pathways, or interaction among pathways, is known in humans or other mammals. Study of these pathways can be done in a variety of ways, at a variety of levels and using a variety of model systems. Planarians offer particular opportunities for studying the neurochemistry of these pathways and in some cases some advantages over mammalian models, because they possess a primitive central nervous system and neurotransmitter systems, but simpler pharmacokinetics related to the absorption, distribution, metabolism and excretion of drugs (singly or in combinations). In addition, they have the advantage of convenience and low cost.

Neurochemical Pathways

Dopamine Pathways

Some of the brain circuits relevant to drug addiction (equated to activation of neurochemical reward pathways) involve dopaminergic pathways, such as the mesolimbic dopamine system.[1-3] The mesolimbic dopamine system extends from dopamine containing cell bodies within the ventral tegmental area (VTA) in brainstem to the nucleus accumbens (NuAcc) (part of the basal ganglia), prefrontal cortex and dorsal striatum and reciprocal projections. An hypothesis is that many abused substances enhance dopamine release in either the nucleus accumbens, the prefrontal cortex, or both.[4] For example, amphetamine, cocaine, ethanol and nicotine all increase the extracellular levels of dopamine in the NuAcc and lesions of mesolimbic dopamine neurons attenuate nicotine self-administration in rats.[5-8] Areas that receive projections from the nucleus accumbens, such as the globus pallidus and amygdala, are also believed to be important. In addition, other monoaminergic nuclei, such as those in the locus coeruleus (norepinephrine-containing cell bodies) and raphe (5-HT containing cell bodies) are also believed to be important.[9] Further, chronic use of drugs leads to disruption of normal homeostatic levels of neurotransmitters and abrupt withdrawal unmasks the compensatory adjustments, which may explain some of the dysphoric aspects of withdrawal syndromes.[10]

Opioid Pathways

The opiates, morphine and codeine and related opioids such as heroin, produce their effects by mimicking endogenous substances such as β-endorphin and Leu- and Met-enkephalin. Opioids activate 7-transmembrane G protein-coupled receptors (mu, delta and kappa).[11] The mu-opioid

*Robert B. Raffa—Department of Pharmaceutical Sciences, Temple University School of Pharmacy, 3307 N. Broad Street, Philadelphia, Pennsylvania, USA.
Email: robert.raffa@temple.edu

Planaria: A Model for Drug Action and Abuse, edited by Robert B. Raffa and Scott M. Rawls.
©2008 Landes Bioscience.

receptor appears to be the most closely associated with drug dependence, which has been linked to the dopamine system.[12] Cocaine produces multiple pharmacologic effects including a local anesthetic action and is a vasoconstrictor (as a consequence of inhibition of neuronal reuptake of norepinephrine).[13] The primary mechanism of action believed to be related to its misuse is the inhibition of the dopamine transporter, which is responsible for the reuptake of dopamine into the presynaptic nerve terminal.[12,14] Inhibition of the dopamine transporter (DA-T) increases the synaptic concentrations of dopamine enabling more activation of DA receptors. This mechanism is supported by experiments on genetically altered mice lacking the DA-T.[15] These mice mimic the behavioral actions of cocaine without receiving the stimulant and show no further changes in behavior after cocaine administration.[14,15] The amphetamines block monoamine neuronal reuptake and enhance their release.[16] Although amphetamines and cocaine raise synaptic concentrations of the three monoamine neurotransmitters norepinephrine (NE), dopamine and 5-HT, selective antagonists for only dopamine block the rewarding effects.[17,18]

Other Neurotransmitter Pathways

Dopamine appears to have a central role particularly in the early stages of initiation of compulsive drug use.[14,19] Noradrenergic system activity is decreased in certain brain regions by the opioids. The compensatory process produces a hyperadrenergic state which might explain some adrenergic-like symptoms during withdrawal and the amelioration of some of these symptoms by drugs such as clonidine which inhibit the release of norepinephrine from the presynaptic nerve terminal.[9] Interactions between the dopamine and norepinephrine systems can occur in the nucleus accumbens and prefrontal cortex, neuroanatomic areas thought to be important in drug abuse.[20] Changes in 5-HT systems are believed to be involved in the changes in appetite, in impulsivity and in the craving that follows abstinence.[9] Alterations in the nicotinic cholinergic receptor system follow chronic nicotine use.[21] In addition, chronic use of drugs alters the levels of endogenous neuropeptides. In some instances the endorphins and the enkephalins are activated, possibly explaining the therapeutic benefit of the opioid antagonist naltrexone in the treatment of alcohol dependence.[22] Chronic drug use also leads to cell molecular adaptations, such as at the level of second messenger transduction systems and protein transcription.[23]

Reward Pathways

Positive reinforcing effects are important for the establishment of a habit or pattern of continued drug use that might lead to drug-seeking behavior.[24] An anatomical pathway in the brain that is involved with reward or reinforcement of specific behaviors was described in 1954 by Olds and Milner.[25] Activation of these neuronal pathways, either electrically or chemically, is reinforcing and can maintain established behaviors, suggesting the concept of an anatomical 'reward' pathway.[26] As reviewed by Wise,[20,27] the mesolimbic dopamine system, including its projections to the nucleus accumbens and local GABAergic afferents, has been most clearly associated with the habit-forming aspects of drugs of abuse. The evidence includes: (i) lesions in the NuAcc attenuate the rewarding effects of cocaine[28,29] and amphetamine,[30] (ii) rats learn to lever-press for microinjection into the NuAcc of amphetamine,[31,32] dopamine[33] and selective dopamine reuptake inhibitors[34] (iii) nACh receptors on dopamine cells are important for nicotine reward[6] and (iv) dopamine levels are elevated in the NuAcc by opioids, nicotine, ethanol and cannabis.[4,5,35,36] According to the extension of these findings to drug abuse, activation of reward pathways reinforces drug-seeking behavior, possibly to a greater extent in individuals with an enhanced sensitivity or responsiveness to activation of the critical brain regions.

Final Common Pathway

Recent research has raised the possibility of a common biochemical mechanism involved in drug abuse. One candidate is the neurotransmitter dopamine and its pathways.[2,3,37-45] Serotonin was also implicated early,[46-48] with specific targets analogous to those of dopamine. The sole involvement of dopamine in addictions has recently come under some question. It was reported recently[49] that rats in which stimulating electrode placement elicited dopamine release in the NuAcc learned to press a lever in order to receive an intracranial electrical stimulation (ICS), but the dopamine release was consistently observed only when the stimulus was applied to an untrained animal and not during ICS. The authors conclude that dopamine might be "...a neural substrate for novelty or reward expectation rather than reward itself." They cite research in support of this position which shows that dopamine neurons in the substantia nigra of monkeys increase their firing rate when an appetitive reward is delivered in an unpredictable way, but not if a conditioned stimulus (tone) precedes the reward. In the latter case, the rate increases not to the reward, but rather to the tone.[50]

Use Schedules

Equally important to the overall problem are the factors that promote continued drug use, 'binge' patterns of use and relapse. In the case of maintenance, avoidance of negative reinforcement (dysphoric affective and physical withdrawal symptoms) plays an important role.[51] As reviewed by Kreek and Koob,[52] chronic drug use alters several neurotransmitters (including dopamine) and other biochemical, systems. Recent work has also suggested that protracted abstinence can change the 'set point' for hedonic processing or relieving physical or mental discomfort. A prolonged reward dysregulation occurs for all major drugs of abuse. Chronic drug use also results in the recruitment of systems, perhaps involving dynorphin, neuropeptide FF or orphanin FQ, which counteract the changes induced by the abused drugs.[52] Hence, cessation of drug-taking results in physical and affective motivational impetus for reinstatement (relapse). Drug-seeking and drug-craving often persist despite long periods of abstinence, the consequence of long-term neuroadaptations in brain reward or anti-reward systems.[53]

Poly-Drug Abuse

Various pharmacologic substances are abused. Under real-world conditions, many drug abusers engage in 'poly-drug' abuse—i.e., the abuse of more than one controlled substance, often simultaneously. Because of the well-known crosstalk among neurotransmitter and receptor systems, it is reasonable to hypothesize that the development of physical dependence in such individuals—and the appropriate treatment strategies—might differ from that of individuals who abuse only a single drug. Although the individual mechanisms of action are somewhat well known for some of these substances, it is now recognized that many drug abusers use multiple drugs, often in combination ('poly-drug' abuse). Each of the abused drugs modulate biochemical process(es) within the central nervous system and cause homeostatic dysregulation, alter the hedonic 'set-point' and activate the biochemical 'switch' that leads to more chronic and detrimental drug abuse, craving and relapse.[54-56] Through extensive investigation the biochemistry of abused drugs is becoming clearer. Many neurotransmitter systems are altered during chronic drug (ab)use. Since the abused drugs alter multiple brain pathways, study of the interaction among the pathways is critically important. In the relatively few studies that have been conducted, specific drug interactions have been identified between dopamine and opioid (rats),[57-59] dopamine and 5-HT (rats),[60,61] opioid and cocaine (rats and rhesus monkeys).[62-64]

Planarians as Model

Although monkey and rodent models are valuable for single-drug abuse studies, they suffer from limitations that impede their usefulness for poly-drug abuse studies. The principal problem results from possible pharmacokinetic interaction. In the case of individual drugs, the dose can be used as an estimate of the concentration at the target site, despite the fact that the (in)accuracy of the estimate is highly dependent on pharmacokinetic factors, such as absorption, distribution (blood-brain barrier), metabolism, etc. Although the drug concentration at the target site is not known, it is correlated to the dose. However, in the case of multiple drugs, the PK of one can alter the PK of the other, thus altering the correlation between dose and target concentration within the same experiment.

Planarians offer an attractive alternative because of their very extensive history of use in a host of anatomic, physiologic, psychologic (learning and memory) and biochemical studies. Planarians have been a staple model for behavioral and biochemical research for decades. A substantial amount of evidence has shown that planarians respond to dopaminergic agonists, antagonists or neuronal reuptake inhibitors with characteristic behaviors or changes in locomotor activity (motility) (e.g., refs. 65-69). Changes in second messenger levels, such as increase in cAMP induced by dopamine agonists,[69] also suggest that the changes in planarian locomotor activity are mediated via dopamine receptors. In addition, drugs of abuse such as amphetamine and cocaine induce 'screw-like hyperkinesias' (SLH) or 'C-like' position (CLP) (e.g., refs. 66,68,69).

Histochemical and other techniques have revealed the presence of monoamine containing neurons, catecholamines and 5-HT in planarians.[65,70-72] That sympathomimetic agents produce easily measurable biochemical responses in planarians has also been known for some time. Dopamine is the most potent stimulatory agent (dopamine > epinephrine > isoproterenol > ephedrine).[73] That some interaction between neurotransmitter systems occurred was reported as early as 1975.[66] Moreover, as in mammals, dopamine agonists stimulate cAMP formation in planarians in a manner related to mu opioid receptors.[68,74] An opioid-dopamine behavioral interaction has been described recently,[75] following the identification of an enkephalinergic system in planarians by means of radioimmunological and immunocytochemical techniques.[74] Also, a planarian 5-HT receptor was recently identified by PCR that binds LSD.[76]

Establishment of the Planarian Model

The study of development of physical dependence and withdrawal in planarians benefits by having a quantifiable metric. An early version of the metric described below was developed by Needleman;[77] others certainly must have been envisioned or utilized.

The pLMA Metric

A substantial amount of evidence had shown that planarians respond to dopaminergic agonists, antagonists or neuronal reuptake inhibitors with characteristic behaviors or changes in locomotor activity (motility) (e.g., refs. 65-69). Changes in second messenger levels, such as increase in cAMP levels induced by dopamine agonists,[69] also suggested that the changes in planarian locomotor activity are mediated via dopamine receptors. In addition, abused drugs such as amphetamine and cocaine had been previously shown to induce what has been described as 'screw-like hyperkinesias' (SLH) or 'C-like' position (CLP) (e.g., refs. 66,68,69). These characterizations of changes in planarian behavioral patterns were mostly qualitative in nature (e.g., induction of SLH or CLP). In addition, it was not entirely clear that the effect of dopamine receptor ligands was related to stereospecific receptor mechanisms. A convenient quantitative endpoint of planarian locomotor activity was sought.[78] An overview of the devised metric is described below.

Black or brown planarians (*Dugesia dorotocephala, s.l.*) are placed individually into a clear plastic petri dish (14 cm diameter; other sizes also work) containing room-temperature (19°C) tap water (treated with AmQuel® water conditioner) located over graph paper consisting of gridlines (square-pattern) spaced 0.5 cm apart. Each planarian locomotor velocity (pLMV) is defined and quantified as the number of gridlines crossed or recrossed by each planarian per minute over an observation period (5 minutes is usually sufficient) conducted in a well-lit room. The pLMV is expressed as the mean (± S.E.M.) of the cumulative number of gridlines crossed by each planarian per minute. Relative potency is calculated from linear regression analysis of individual dose-response curves.

Under these conditions, planarians display a characteristic and stable locomotor velocity over the course of the observation period. The average velocity typically falls within a relatively narrow range of about 14-16 gridline crossings per minute. The planarians sometimes display a slightly greater locomotor velocity during the first 1-2 min of observation, but the velocity remains essentially constant over the remainder of the period, as shown in Figure 1. Thus, pLMV is a robust

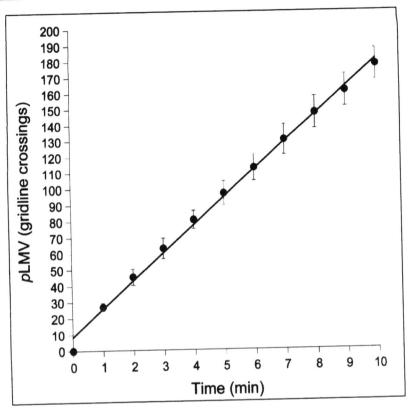

Figure 1. Planarian spontaneous locomotor activity (*p*LMA) expressed as the mean ± S.D. of the cumulative number of gridlines crossed over a 10 min observation. The consistency is shown by the small error (N = 5 planarians). Data supplied by Amrita Pandya, PharmD student, Temple University School of Pharmacy.

and quantifiable endpoint that provides a stable baseline behavior against which to compare the action of pharmacologic test agents.

Stereoselective Response

Dopaminergic ligands had been shown to induce or modify characteristic patterns of motility in planarians. For example, as described above, apomorphine or amphetamine induce 'screw-like hyperkinesias' (SLH) or 'C-like' position (CLP). That the effect is related to dopamine was supported by the identification of monoamine-containing neurons and detection of dopamine in planarians.[65,67] Importantly, the agonist-induced behavioral effects are antagonized by haloperidol.[66,68] Further evidence that dopamine receptors are involved includes: D_1-selective agonists (e.g., CY 208243, SKF 38393) induce SLH and the effect is antagonized by a D_1 (SCH 23390), but not D_2 (sulpiride), antagonist; D_2-selective agonists (e.g., PHNO or lisurgide) induce CLP and the effect is antagonized by a D_2, but not D_1, antagonist; and, dopamine agonist-induced increase in cAMP is antagonized by pretreatment with receptor antagonist.[67,68] To strengthen the evidence for dopamine receptor involvement, the effect of the enantiomers of the selective dopamine D_2 receptor antagonist sulpiride (5-(Aminosulfonyl)-*N*-[(1-ethyl-2-pyrrolidinyl) methyl]-2-methoxy-benzamide) was examined.[79] The measurement of *p*LMV was the same as above, but prior to the measurement, each planarian was placed in room-temperature, treated tap water (AmQuel® water conditioner) (the vehicle) or a solution of S(−)-sulpiride or R(+)-sulpiride $(1 \times 10^{-13}$ to 1×10^{-4} M).

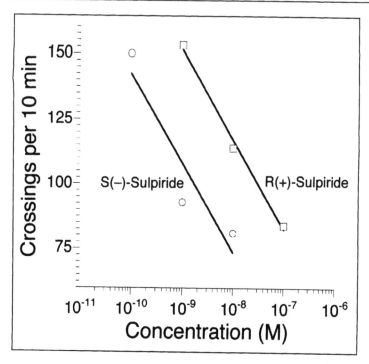

Figure 2. Demonstration of an enantiomer-sensitive drug-induced effect on *p*LMV by the dopamine D_2 receptor antagonist S(–)-sulpiride and its less active enantiomer R(+)-sulpiride.

Planarians exposed to S(–)-sulpiride displayed reduced *p*LMV compared to untreated or vehicle-treated controls. The S(–)-sulpiride-induced decrease of *p*LMV was dose-related: crossings per 10 min (C10) was related to S(–)-sulpiride dose according to C10 = –34.5 *log*(M)–202.5. Those planarians that were exposed to the enantiomer of S(–)-sulpiride, i.e., R(+)-sulpiride, displayed a parallel reduction in *p*LMV compared to S(–)-sulpiride-treated planarians, but R(+)-sulpiride was 25-fold less potent than S(–)-sulpiride in producing the effect: C10 was related to R(+)-sulpiride dose according to C10 = –34.5 *log*(M)–159.0 (Fig. 2). In separate tests, planarians exposed to the dopamine D_1 receptor antagonist R(+)-SCH 23390 (7-chloro-8-hydroxy-3-methyl-5-phenyl-2,3,4,5-tetrahydro-1*H*-3-benzazepine) responded only to high doses ($\geq 10^{-5}$ M) and equally to its enantiomer S(–)-SCH 23388 (data not shown).

*p*LMV was reduced in a dose-dependent and enantiomer-preferred manner by low concentration ($\geq 1 \times 10^{-10}$ M) of S(–)-sulpiride, but *p*LMV was not reduced by low concentrations ($< 1 \times 10^{-5}$ M) of the enantiomer of S(–)-sulpiride (R(+)-sulpiride). Hence, *p*LMV was sensitive to a dopamine antagonist in a fashion characteristic of the involvement of receptor-mediated transduction: i.e., dose-dependent and stereoselective.

The Effect of Short-Wave, but not Long-Wave, UV Light

To further support the involvement of binding sites, rather than nonspecific action, in the effect of dopamine ligands on *p*LMV, the effect of UV light was examined.[79] UV light-induced relaxation of isolated thoracic aorta previously contracted to steady-state isometric tension by alpha-adrenoceptor agonists is a well-known phenomenon.[80] In prior work described above, it was observed that the D_2 receptor antagonist sulpiride decreases *p*LMV in an enantiomeric-selective ((–)sulpiride >> (+)sulpiride) and dose-dependent manner. It was determined if irradiation by UV light of different wavelengths (energy levels) would disrupt the (−)sulpiride-D_2 bond and attenuate the effect of (−)sulpiride on pLMV. The pLMV was measured as described above. Six groups were tested: (1) untreated planarians; (2) planarians exposed to (−)sulpiride (0.1 µM); (3) planarians exposed to long-wave (366 nm) UV light (UV-L) (placed 5-in above); (4) planarians exposed to short-wave (254 nm) UV light (UV-S) (5-in perpendicular above); (5) the combination of (2) and (3); and (6) the combination of (2) and (4).

Neither UV-L nor UV-S alone had any effect on the pLMV of untreated planarians. (−)Sulpiride (0.1 µM) reduced the slope of the pLMV, consistent with prior findings. The (−)sulpiride-induced decrease of pLMV was attenuated by UV-L (70%) and was attenuated to an even greater extent by the UV-S (84%). Thus, the locomotor activity of the planarians that was suppressed by (−)sulpiride reverted back toward control (untreated) levels when the planarians were exposed to UV light during testing. It seems unlikely that the changes can be explained as toxic effects, since the UV light reverted the pLMV toward normal levels and the effect was greater for high-energy UV light (254 nm = 7.83 × 10-19 J = 4.89 eV) than for low energy UV light (366 nm = 5.43 × 10-19 J = 3.39 eV). The results are more consonant with the view that the UV light either (1) stimulated the release of some as yet unidentified substance in a wavelength-dependent manner, or (2) disrupted the binding or transduction process. The latter view is consistent with a considerable body of evidence from study of UV-induced photorelaxation of rabbit thoracic aorta suggesting that UV light disrupts the drug-receptor bond.[81-84] Subsequent work with amphetamine (unpublished) supports the receptor-related interpretation.

Withdrawal Signs

To be convincing, 'withdrawal' in planarians should be accompanied by withdrawal signs. Indeed, planarians, like mammals, display characteristic signs when they undergo withdrawal.[85] Cocaine-experienced planarians placed into cocaine-free water display behaviors not typically seen in drug-naïve planarians. These behaviors are illustrated in Figure. 3 and were loosely termed: 'HeadBop' ('nodding' movement of its head as the planarian moved forward); 'Squirming' (uncoordinated, 'jerky' movements); 'Clinging' ('scrunching' of the body); 'HeadSwing' (axial rotation of head about long axis 'helicopter' motion—while tail is anchored); 'TailTwist' (tip of body twisted) and 'Corkscrew' (spiral motion around long axis). The occurrence of the behaviors were most pronounced during the first 5-min observation period and declined during the subsequent observation periods.

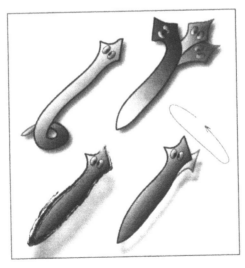

Figure 3. Cartoons of the atypical behaviors displayed by planarians undergoing drug withdrawal.

Conclusions

Planarians provide a convenient, reliable and robust model for investigation of phenomena related to drug action and abuse. They have a centralized nervous system and neurotransmitter systems that utilize the same neurotransmitters and possibly 2nd-messenger systems, as mammals (albeit in a more primitive manner). They have relatively advanced behaviors, such as learning and memory and they display a withdrawal 'syndrome' during abstinence-induced or antagonist-induced drug withdrawal.

References

1. Koob GF. Drugs of abuse: Anatomy, pharmacology and function of reward pathways. Trends Pharmacol Sci 1992; 13:177-184.
2. Altman J. A biological view of drug abuse. Molec Med Today 1996; 2:237-241.
3. Nash JM. Addicted. Time 1997; 149:69-76.
4. Di Chiara G, Imperato A. Drugs abused by humans preferentially increase synaptic dopamine concentrations in the mesolimbic system of freely moving rats. Proc Natl Acad Sci (USA) 1998; 85:5274-5278.
5. Imperato A, Mulas A, Di Chiara G. Nicotine preferentially stimulates dopamine release in the limbic system of freely moving rats. Eur J Pharmacol 1986; 132:337-338.
6. Corrigall WA, Franklin KB, Coen KM et al. The mesolimbic dopaminergic system is implicated in the reinforcing effects of nicotine. Psychopharmacol 1992; 107:285-289.
7. Benwell ME, Balfour DJ, Khadra LF. Studies on the influence of nicotine infusions on mesolimbic dopamine and locomotor responses to nicotine. Clin Invest 1994; 72:233-239.
8. Pontieri FE, Tanda G, Orzi F et al. Effects of nicotine on the nucleus accumbens and similarity to those of addictive drugs. Nature 1996; 382:255-257.
9. Nutt DJ. Addiction: Brain mechanisms and their treatment implications. Lancet 1996; 347:31-36.
10. Gawin FH. Cocaine addiction: Psychology and neurophysiology. Science 1991; 251:1580-1586.
11. Akil H, Owens C, Gutstein H et al. Endogenous opioids: Overview and current issues. Drug Alcohol Depend 1998; 51:127-140.
12. Gatley SJ, Volkow ND. Addiction and imaging of the living human brain. Drug Alcohol Depend 1998; 51:97-108.
13. Karch SB. The Pathology of Drug Abuse. 2nd ed. Boca Raton: CRC Press, 1996:1-175.
14. Amara SG, Sonders MS. Neurotransmitter transporters as molecular targets for addictive drugs. Drug Alcohol Depend 1998; 51:87-96.
15. Giros B, Jaber M, Jones SR et al. Hyperlocomotion and indifference to cocaine and amphetamine in mice lacking the dopamine transporter. Nature 1996; 379:606-612.
16. Heikkila RE, Orlansky H, Mytilineou C et al. Amphetamine: Evaluation of d- and l-isomers as releasing agents and uptake inhibitors for [³H]dopamine and ³H-norepinephrine in slices of rat neostriatum and cerebral cortex. J Pharmacol Exp Ther 1975; 194:47-56.
17. de Wit H, Wise RA. Blockade of cocaine reinforcement in rats with the dopamine receptor blocker pimozide but not with the noradrenergic blockers phentolamine or phenoxybenzamine. Canad J Psychol 1977; 31:195-203.
18. Lacosta S, Roberts DCS. MDL 72222, ketanserin and methysergide pretreatments fail to alter breaking points on a progressive ratio schedule reinforced by intravenous cocaine. Pharmacol Biochem Behavior 1993; 44:161-165.
19. Hubner CB, Moreton JE. Effects of selective D_1 and D_2 dopamine antagonists on cocaine self-administration in the rat. Psychopharmacol 1991; 105:151-156.
20. Wise RA. Drug activation of brain reward pathways. Drug Alcohol Depend 1998; 51:13-22.
21. Picciotto MR. Common aspects of the action of nicotine and other drugs of abuse. Drug Alcohol Depend 1998; 51:165-172.
22. Volpicelli JR, Alterman AI, Hayashida M et al. Naltrexone in the treatment of alcohol dependence. Arch Genl Psychiatr 1992; 49:876-880.
23. Kosten TR, Hollister LE. Drugs of abuse. In: Katzung BG. Basic and Clinical Pharmacology. 7th ed. Stamford CT: Appleton and Lange, 1998:516-531.
24. Wise RA. The neurobiology of craving: Implications for understanding and treatment of addiction. J Abnorm Psychol 1988; 97:118-132.
25. Olds J, Milner PM. Positive reinforcement produced by electrical stimulation of septal area and other regions of rat brain. J Comp Physiol Psychol 1954; 47:419-427.
26. Wise RA. Action of drugs of abuse on brain reward systems. Pharmacol Biochem Behav 1980; 13(Suppl. 1): 213-223.
27. Wise RA. Addictive drugs and brain stimulation reward. Ann Rev Neurosci 1996; 19:319-340.

28. Roberts DCS, Koob GF, Klonoff P et al. Extinction and recovery of cocaine self-administration following 6-OHDA lesions of the nucleus accumbens. Pharmacol Biochem Behav 1980; 12:781-787.
29. Roberts DCS, Corcoran ME, Fibiger HC. On the role of ascending catecholaminergic systems in intravenous self-administration of cocaine. Pharmacol Biochem Behav 1977; 6:615-620.
30. Lyness WH, Friedle NM, Moore KE. Destruction of dopaminergic nerve terminals in nucleus accumbens: effect on D-amphetamine self-administration. Pharmacol Biochem Behav 1979; 11:553-556.
31. Hoebel BG, Monaco AP, Hernandez L et al. Self-injection of amphetamine directly into the brain. Psychopharmacol 1983; 81:158-163.
32. Yokel RA, Wise RA. Increased lever-pressing for amphetamine after pimozide in rats: Implications for a dopamine theory of reward. Science 1975; 187:547-549.
33. Dworkin SI, Goeders NE, Smith JE. The reinforcing and rate effects of intracranial dopamine administration. NIDA Res Monograph 1986; 67:242-248.
34. Carlezon WA Jr, Devine DP, Wise RA. Habit-forming actions of nomifensine in nucleus accumbens. Psychopharmacol 1995; 122:194-197.
35. Ng Cheong Ton JM, Gerhardt GA, Friedemann M et al. The effects of Δ^9-tetrahydrocannabinol on potassium-evoked release of dopamine in the rat caudate nucleus: An in vivo electrochemical and in vivo dialysis study. Brain Res 1988; 451:59-68.
36. Devine DP, Leone P, Pocock D et al. Differential involvement of ventral tegmental mu, delta and kappa opioid receptors in modulation of basal mesolimbic dopamine release: In vivo microdialysis studies. J Pharmacol Exper Ther 1993; 266:1236-1246.
37. Balter M. New clues to brain dopamine control, cocaine addiction. Science 1996; 271:909.
38. Uhl GR, Vandenbergh DJ, Miner LL. Knockout mice and dirty drugs. Current Biol 1996; 6:935-936.
39. Caron MG. Knockout of the vesicular monoamine transporter 2 gene results in neonatal death and supersensitivity to cocaine and amphetamine. Neuron 1997; 19:1285-1296.
40. Takahashi N, Miner LL, Sora I et al. VMAT2 knockout mice: Heterozygotes display reduced amphetamine-conditioned reward, enhanced amphetamine locomotion and enhanced MPTP toxicity. Proc Natl Acad Sc (USA) 1997; 94:9938-9943.
41. Jones SR, Gainetdinov RR, Wightman RM et al. Mechanisms of amphetamine action revealed in mice lacking the dopamine transporter. J Neurosci 1998; 18:1979-1986.
42. Rocha BA, Fumagalli F, Gainetdinov RR et al. Cocaine self-administration in dopamine-transporter knockout mice. Nature Neurosci 1998; 1:132-137.
43. Pich EM, Epping-Jordan MP. Transgenic mice in drug dependence research. Annals Med 1998; 30:390-396.
44. Sora I, Wichems C, Takahashi N et al. Cocaine reward models: Conditioned place preference can be established in dopamine- and in serotonin-transporter knockout mice. Proc Natl Acad Sci USA 1998; 95:7699-7704.
45. Robbins TW, Everitt BJ. Drug addiction: Bad habits add up. Nature 1999; 398:567-570.
46. Kleven MS, Woolverton WL. Effects of three monoamine uptake inhibitors on behavior maintained by cocaine or food presentation in rhesus monkeys. Drug Alcohol Depend 1993; 31:149-158.
47. Spealman RD. Modification of behavioral effects of cocaine by selective serotonin and dopamine uptake inhibitors in squirrel monkeys. Psychopharmacol 1993; 112:93-99.
48. Walsh SL, Preston KL, Sullivan JT et al. Fluoxetine alters the effects of intravenous cocaine in humans. J Clin Psychopharmacol 1994; 14:396-407.
49. Garris PA, Kilpatrick M, Bunin MA et al. Dissociation of dopamine release in the nucleus accumbens from intracranial self-stimulation. Nature 1999; 398:67-69.
50. Mirenowicz J, Schultz W. Preferential activation of midbrain dopamine neurons by appetitive rather than aversive stimuli. Nature 1996; 379:449-451.
51. Russell MAH. What is dependence? In: Edwards G, Russell MAH, Hawks D et al, eds. Drugs and Drug Dependence. Lexington: Lexington Books, 1976:182-187.
52. Kreek MJ, Koob GF. Drug dependence: stress and dysregulation of brain reward pathways. Drug Alcohol Depend 1998; 51:23-47.
53. Self DW, Nestler EJ. Relapse to drug-seeking: neural and molecular mechanisms. Drug Alcohol Depend 1998; 51:49-60.
54. Koob GF. Neurochemical explanations for addiction. Hosp Pract 1997; (Special Report):12-14.
55. Self DW. Neurobiological adaptations to drug use. Hosp Pract 1997; (Special Report):5-9.
56. Ahmed SH, Koob GF. Transition from moderate to excessive drug intake: Change in hedonic set point. Science 1998; 282:298-300.
57. Turchan J, Przewlocka B, Lason W et al. Effects of repeated psychostimulant administration on the prodynorphin system activity and kappa opioid receptor density in the rat brain. Neurosci 1998; 85:1051-1059.
58. Woolfolk DR, Holtzman SG. The effects of opioid receptor antagonism on the discriminative stimulus effects of cocaine and d-amphetamine in the rat. Behav Pharmacol 1996; 7:779-787.

59. Collins SL, D'Addario C, Izenwasser S. Effects of kappa-opioid receptor agonists on long-term cocaine use and dopamine neurotransmission. Eur J Pharmacol 2001; 426:25-34.
60. Sasaki-Adams DM, Kelley AE. Serotonin-dopamine interactions in the control of conditioned reinforcement and motor behavior. Neuropsychopharmacol 2001; 25:440-452.
61. Filip M, Nowak E, Papla I et al. Role of 5-hydroxytryptamine$_{1B}$ receptors and 5-hydr-oxytryptamine uptake inhibition in the cocaine-evoked discriminative stimulus effects in rats. J Physiol Pharmacol 2001; 52:249-263.
62. Sharpe LG, Pilotte NS, Shippenberg TS et al. Autoradiographic evidence that prolonged withdrawal from intermittent cocaine reduces mu-opioid receptor expression in limbic regions of the rat brain. Synapse 2000; 37:292-297.
63. Yuferov V, Zhou Y, LaForge KS et al. Elevation of guinea pig brain preprodynorphin mRNA expression and hypothalamic-pituitary-adrenal axis activity by "binge" pattern cocaine administration. Brain Res Bull 2001; 55:65-70.
64. Wang NS, Brown VL, Grabowski J et al. Reinforcement by orally delivered methadone, cocaine and methadone-cocaine combinations in rhesus monkeys: Are the combinations better reinforcers? Psychopharmacol 2001; 156:63-72.
65. Welsh JH, Williams LD. Monoamine containing neurons in Planaria. J Comp Neurol 1970; 138:103-116.
66. Carolei A, Margotta V, Palladini G. Proposal of a new model with dopaminergic-cholinergic interactions for neuropharmacological investigations. Neuropsychobiol 1975; 1:355-364.
67. Algeri S, Carolei A, Ferretti P et al. Effects of dopaminergic agents on monoamine levels and motor behaviour in Planaria. Comp Biochem Physiol C 1983; 74:27-29.
68. Venturini G, Stocchi F, Margotta V et al. A pharmacological study of dopaminergic receptor in planaria. Neuropharmacol 1989; 28:1377-1382.
69. Palladini G, Ruggieri S, Stocchi F et al. A pharmacological study of cocaine activity in planaria. Comp Biochem Physiology C 1996; 115:41-45.
70. Villar D, Schaeffer DJ. Morphogenetic action of neurotransmitters on regenerating planarians—A review. Biomed Environ Sci 1993; 6:327-347.
71. Welsh JH. Catecholamines in the invertebrates. In: Blaschko H, Muschall F, eds. Platelminthes (Flatworms) Catecholamines. Berlin: Springer, 1972:87-88.
72. Welsh JH, Moorhead M. The quantitative distribution of 5-hydroxytryptamine in the invertebrates, especially in their nervous systems. J Neurochem 1960; 6:146-169.
73. Csaba G, Kádár M. The effect of sympathicomimetic agents on carbohydrate metabolism of Planaria. Acta Physiol Acad Sci Hungar Tomas 1979; 53:323-326.
74. Venturini G, Carolei A, Palladini G et al. Radioimmunological and immuno-cytochemical demonstration of Met-enkephalin in planaria. Comp Biochem Physiol C 1983; 74:23-25.
75. Passarelli F, Merante A, Pontieri FE et al. Opioid-dopamine interaction in planaria: A behavioral study. Comp Biochem Physiol C 1999; 124:51-55.
76. Saitoh O, Yuruzuma E, Nakata H. Identification of planarian serotonin receptor by ligand binding and PCR studies. NeuroReport 1996; 8:172-178.
77. Needleman HL. Tolerance and dependence in the planarian after continuous exposure to morphine. Nature 1967; 215:784-785.
78. Raffa RB, Holland LJ, Schulingkamp RJ. Quantitative assessment of dopamine D2 antagonist activity using invertebrate (Planaria) locomotion as a functional endpoint. J Pharmacol Toxicol Meth 2001; 45:223-226.
79. Raffa RB, Valdez JM, Holland LJ et al. Energy-dependent UV light-induced disruption of (−)sulpiride antagonism of dopamine. Eur J Pharmacol 2000; 406:R11-12.
80. Furchgott RF, Sleator W Jr, McCaman MW et al. Relaxation of arterial strips by light and the influence of drugs on this photodynamic effect. J Pharmacol Exp Ther 1955; 113:22-23.
81. Tallarida RJ, Sevy RW, Harakal C et al. Characteristics of photorelaxation in vascular smooth muscle: evidence supporting the hypothesis of drug-receptor equilibrium disturbance. IEEE Trans Biomed Eng 1975; 22:493-501.
82. Jacob LS, Tallarida RJ. Further studies on the action of ultraviolet light on vascular smooth muscle: effect of partial irreversible receptor blockade. Arch Intl Pharmacodyn Ther 1977; 225:166-176.
83. Tallarida RJ, Laskin OL, Jacob LS. Perturbation of drug receptor equilibrium in the presence of competitive blocking agents. J Theo Biol 1979; 61:211-219.
84. Raffa RB, Robinson MJ, Tallarida RJ. Ultraviolet light-induced photorelaxation of agonist-contracted rabbit aorta: further characterization and the estimation of drug-receptor rate constants. Drug Devel Res 1985; 5:359-369.
85. Raffa RB, Desai P. Description and quantification of cocaine withdrawal signs in Planaria. Brain Res 2005; 1032:200-202.

CHAPTER 9

Physical Dependence and Withdrawal in Planarians

*Robert B. Raffa

Abstract

A quantifiable feature of drug abuse in animal models is the development of physical dependence and its expression during withdrawal (either abstinence- or antagonist-induced). Withdrawal has sometimes been proposed to be important to understanding drug 'craving'. Planarians are a convenient model for the study of physical dependence and withdrawal. Physical dependence develops rapidly in planarians and planarians display a withdrawal syndrome that consists of atypical behaviors that are not observed during normal planarian activity (see Chapter 8). Withdrawal is often more easily and reproducibly demonstrated in planarians than in mammals for substances that display mild withdrawal in mammals. It is also more convenient to study and rigorously quantify withdrawal from multiple drugs in combination (as discussed in Chapter 10). The pharmacology of planarians appears to be quite similar to that of mammals—with minor exceptions that are indicative of more primitive receptor or 2nd-messenger transduction systems.

Introduction

Physical dependence is a physiological process that develops during extended administration of almost any drug. It is often viewed as the body's compensatory response to drug stimulus and an effort to regain predrug homeostatic conditions. It is difficult to measure physical dependence directly, but it can be measured as it is manifested during rapid removal (withdrawal) of the drug, because the body's compensatory response is now unopposed by the drug-induced stimulus.

One of the earliest, if not the earliest, example of the use of a planarian model to study physical dependence and withdrawal is that of Needleman.[1] Needleman reasoned that since prior studies suggested planarians could be conditioned,[2-4] they could also be "addicted". He established a model not too dissimilar from those currently used. An acclimated planarian was exposed to a localized bright light and the time taken by the planarian to move to the perimeter of the observation chamber (12-in diameter plastic laboratory dish) was recorded. Dependence was defined as the deterioration of swimming performance when drug (morphine) was withdrawn. Using this methodology, withdrawal was demonstrated for planarians placed in water following 43 days exposure to morphine.

This and other prior reports of drug-induced effects in planarians (e.g., refs. 5-8) prompted us to establish a quantifiable metric of withdrawal in this species. We used similar methodology as a model to investigate the development of physical dependence and withdrawal in planarians for several drug categories, as described below. The general methodology of all of the studies was as follows. Planarians (*Dugesia dorotocephala*) were purchased from a commercial source (the

*Robert B. Raffa—Department of Pharmaceutical Sciences, Temple University School of Pharmacy, 3307 N. Broad Street, Philadelphia, Pennsylvania, USA.
Email: robert.raffa@temple.edu

Planaria: A Model for Drug Action and Abuse, edited by Robert B. Raffa and Scott M. Rawls.
©2008 Landes Bioscience.

Carolina Biological Supply Co., Burlington, NC). They were acclimated to laboratory conditions and were used (once) within 72 h (typically 5-8 planarians per group). Controlled substances were obtained from NIDA (National Institute on Drug Abuse); other chemicals were purchased from commercial sources. All were dissolved in tap water or appropriate vehicle (in which case a vehicle-alone group was tested). For testing, planarians were pretreated for 1 h individually in 1-2 mL of drug or vehicle and were then placed individually into a clear plastic Petri observation dish (14 cm diameter in most of the cases, but dishes of smaller diameter also work well) containing room-temperature water (at pH = 7.0). The dish was placed over standard graphing paper (the gridlines were spaced 0.5 cm apart). Spontaneous planarian locomotor velocity (pLMV) was measured for each planarian by counting the number of gridlines that each crossed or recrossed per minute over the observation period (typically 5-10 min). Typically, at least three groups were examined: (i) planarians pretreated in water or vehicle, then tested in water or vehicle—as a control for exposure to vehicle; (ii) planarians pretreated in test drug and then tested in water—to check for abstinence-induced withdrawal; and (iii) planarians pretreated in test drug then tested in the same concentration of test drug—as a control that the observed change in pLMV in (iii) above only occurred during the absence of the test drug, not during its continued presence. In other protocols, other test drugs or antagonists were present during the pretreatment or test phases to demonstrate, for example, drug interactions or antagonist-induced withdrawal. The withdrawal signs in planarians consist of:[9] 'HeadBop' (a 'nodding' movement of head accompanying forward movement); 'Squirming' (uncoordinated, 'jerky' movements); 'HeadSwing' (axial rotation of the head about the long axis while tail is anchored); 'TailTwist' (tip of body twisted) and 'Corkscrew' (a spiral motion around the long axis). These behaviors are most pronounced during the first 5-minutes of abstinence and then decline during subsequent observation periods, indicating the temporary nature of the effects. It has been consistently found (e.g., refs. 10-12) that drug-naïve planarians display a nearly constant pLMV of about 13-18 gridlines per minute when tested in water. This results in an essentially linear relationship between pLMV and observation time throughout a 5- or 10-minute observation period (Fig. 1) and provides a stable and sensitive background against which to measure drug-induced alterations, such as occur during abstinence-induced or antagonist precipitated withdrawal.

Opioids

Physical Dependence and Withdrawal

The opioid withdrawal syndrome is extensively documented and characterized in humans and in mammals. In humans, opioid withdrawal signs and symptoms typically appear about a half day after discontinuation of long-term excessive use and persist for a few days or a week. The withdrawal syndrome usually includes one or more of the following: "abdominal pain, chills with gooseflesh (i.e., 'going cold turkey'), diarrhea, fever, irritability, muscle aches and jerks (i.e., 'kicking the habit'), running eyes and nose, sleeplessness, sweating, vomiting, worrying and yawning".[13] In mammals, such as rats, the withdrawal syndrome includes hyperactivity, diarrhea and 'wet-dog shakes'.

Planarians are not very responsive to selective μ or δ opioid receptor agonists, but do respond to κ opioid receptor agonists.[14] Pharmacologic evidence of a κ-opioid receptor-mediated abstinence-induced and antagonist-induced withdrawal phenomenon in planarians has been reported.[12] Drug-naïve planarians displayed characteristically relatively constant (nearly linear) pLMV when tested in water, but planarians that were exposed to the κ-opioid receptor agonist U-50,488H (trans-(±)-3,4-dichloro-N-methyl-N-(2-[1-pyrrolidinyl]cyclohexyl)-benzeneacetamide) tested in water displayed a dose-related reduction in pLMV, indicative of withdrawal, since neither U-50,488H-experienced planarians tested in U-50,488H-containing water or naïve planarians tested in U-50,488H-containing water displayed any change in pLMV compared to U-50,488H-naïve planarians (Fig. 2A). A change in osmolarity or pH, rather than withdrawal, was not the explanation of the observed effect, because neither saline during treatment nor during testing had any effect and the pH of solutions of U-50,488H were not different from water. The

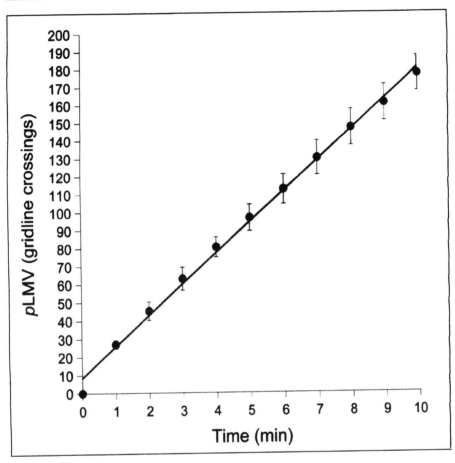

Figure 1. Spontaneous planarian locomotor velocity (*p*LMV) of untreated planarians tested in water. *p*LMV is measured as the number of gridlines crossed each minute. The graphed points are the mean ± S.D. of N = 5 planarians. Data supplied by Amrita Pandya, PharmD student, Temple University School of Pharmacy.

constancy of the *p*LMV slope of U-50,488H-exposed planarians tested in water and reversibility of the effect argues against toxicity as an explanation of the effect.

The enantiomeric selectivity of the withdrawal observed in this study was demonstrated because withdrawal was significantly greater following exposure to U-50,488H, which is the pharmacologically more active enantiomer, than to (+)U-50,488 ([*1R,2R*]U-50,488), which is the pharmacologically less active enantiomer (Fig. 2B). The enantiomeric selectivity suggested the involvement of a (κ-opioid receptor) receptor-mediated pathway. As a more direct test of the involvement of opioid receptors in the effect of U-50,488H, planarians were co-incubated with the opioid antagonist naloxone during exposure to U-50,488H. Planarians that were co-incubated with naloxone plus U-50,488H during pretreatment and then tested in water displayed *p*LMV that was not different from drug-naïve planarians (i.e., there was no abstinence-induced withdrawal). In addition, antagonist-precipitated withdrawal was shown when naloxone added to the test dish produced dose-related decrease in *p*LMV in U-50,488H-experienced planarians (Fig. 2C). As further demonstration of κ-opioid receptor involvement, planarians that were co-incubated with the κ-opioid receptor antagonist *nor*-BNI (*nor*-Binaltorphimine) during exposure to U-50,488H displayed *p*LMV that was not different from naïve planarians.

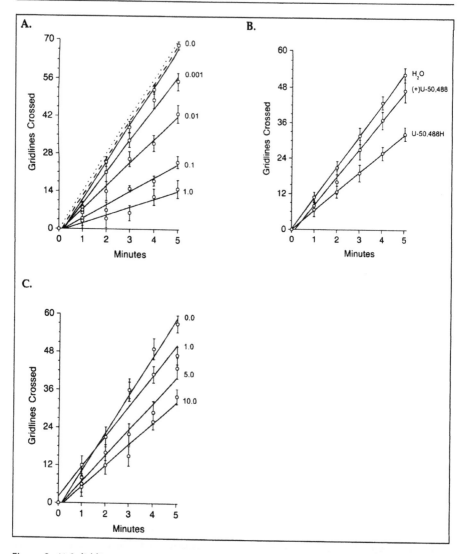

Figure 2. A) Solid lines: pLMV (mean ± S.D.) of planarians previously exposed to U-50,488H (concentration in μM indicated) for 1 h and then tested in water. Dotted line: U-50,488H-naïve animals tested in water. Dashed line: U-50,488H-exposed animals tested in the same concentration of U-50,488H. N = 3 planarians each group. B) Comparison of the effect of (+)U-50,488 and racemic U-50,488H. N = 5 planarians per group. C) pLMV (means ± S.D.) of planarians exposed to U-50,488H for 1 h then tested in U-50,488H plus naloxone at the concentration (μM) indicated. N = 5 planarians per group. Reprinted with permission from reference 12.

In summary, abstinence-induced withdrawal from U-50,488H was elicited when U-50,488H-experienced planarians where placed into U-50,488H-free water, but the effect was not elicited when U-50,488H-experienced planarians were placed into water containing U-50,488H. Evidence for the involvement of opioid receptors in the mediation of the observed withdrawal was provided from two different approaches: (i) antagonism of the abstinence-induced U-50,488H withdrawal by the opioid antagonist naloxone and (ii) dose-related naloxone-precipitated withdrawal of U-50,488H-experienced planarians. Evidence for the more specific involvement of κ subtype of opioid receptors was obtained from two approaches: (i) the antagonism of abstinence-induced U-50,488H withdrawal by the κ-selective opioid antagonist *nor*-BNI (which had no effect of its own on *p*LMV) and (ii) the more than 2-fold greater effect of racemic U-50,488H than the less active enantiomer (+)U-50,488. These findings offered pharmacologic evidence of a receptor-mediated opioid withdrawal phenomenon in the planarian model.

Schild (apparent pA_2) Analysis

According to conventional drug-receptor theory, a ligand (L) binds with weak chemical and physical bonds to its receptor (R) in a reversible interaction to form a ligand-receptor complex (LR) according to

$$L + R \Leftrightarrow LR.$$

The interaction is quantitatively characterized by a dissociation constant, which is the reciprocal of the standard chemical equilibrium constant. The dissociation constant is commonly designated K_A when L is an agonist at the receptor and K_B when L is an antagonist at the receptor. The K_B of a 'surmountable' ('reversible' or 'competitive') antagonist can be measured by relating the shift of an agonist's dose-response curve to the concentration of antagonist required to produce the observed shift. The procedure is often termed 'pA_2 analysis'. [15] If at any level of effect the concentrations of agonist are denoted A in the absence of antagonist and A' in the presence of concentration of antagonist B, then applying the law of mass action leads to

$$(A'/A) - 1 = B / K_B,$$

where A'/A is the agonist dose-ratio for each concentration B and K_B is the dissociation constant of the surmountable antagonist. The pA_2 is defined as the negative common logarithm of the concentration B that produces a dose-ratio equal to 2 (i.e., 2-fold shift of the agonist dose-response curve). [16] The above equation is equivalent to

$$\log(A'/A - 1) = (\log B) - \log K_B.$$

Thus plotting $\log(A'/A - 1)$ against $-\log B$ (molar) produces a 'Schild-plot'. The results plot as a straight line for receptor-mediated effects when the interaction between agonist and antagonist for the same receptor is competitive. In this case, the x-intercept yields the pA_2 of the antagonist and, if the slope $= -1$, the K_B of the antagonist. [15]

We applied pA_2 analysis to the planarian model, [17] specifically related to the κ-opioid response. This was done by constructing a series of rightward, parallel shifts of the κ-opioid receptor agonist U-50,488H dose-response curve by the selective κ-opioid receptor surmountable antagonist *nor*-BNI (Fig. 3). The pA_2 value calculated from these data was 6.05, which is relatively close to the value of 6.5 reported for *nor*-BNI in a mouse test. [18] The lower affinity in planarians compared to a mammal (6.05 vs 6.5) is consistent with previous observations of invertebrates. [19] These results add additional pharmacologic evidence that U-50,488H and *nor*-BNI act as agonist and antagonist, respectively, at a common receptor—which is a presumptive planarian equivalent of the mammalian κ-opioid receptor.

Cocaine

Less obvious or consistent effects are experienced during withdrawal from cocaine than are commonly experienced during withdrawal from opioids. It is reported that when long-term cocaine use is discontinued, "apathy, depression, dysphoria, fatigue, headaches, muscle cramps, psychomotor retardation, sleep disturbances (e.g., insomnia) and vivid unpleasant dreams occur". [20] Withdrawal from cocaine is demonstrated easily in planarians.

When cocaine-experienced planarians where placed into cocaine-free water, an abstinence-induced cocaine withdrawal was elicited, manifested and quantified as a decrease in *p*LMV and was not elicited when cocaine-exposed planarians were placed into cocaine-containing water. [10] The magnitude of the effect was dose-dependently related to prior cocaine exposure

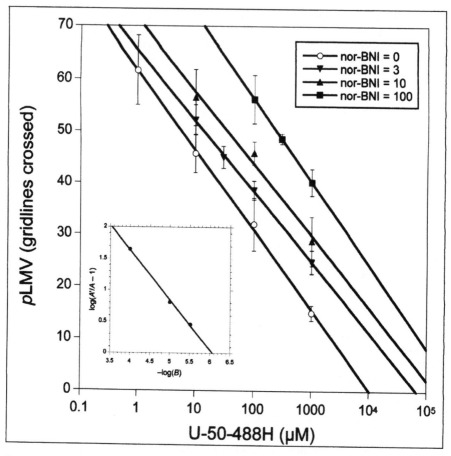

Figure 3. Rightward, parallel shifts of the U-50,488H dose-response curve (*p*LMV endpoint) produced by three doses of nor-BNI (3, 10 and 100 μM). N = 5-10 planarians per point. Inset: 'Schild plot' of the data: shift ratios of the agonist A) dose-response curves plotted against the corresponding antagonist B) concentration. Reprinted with permission from reference 17.

(Fig. 4A). In subsequent work, it was shown that cocaine withdrawal in planarians is blocked by D-, but not L-, glucose (Fig. 4B).[21]

Benzodiazepines

Withdrawal signs and symptoms have been described following the abrupt discontinuation of long-term use of benzodiazepine anxiolytics.[22] However, physical dependence generally develops slowly in humans and abstinence-induced withdrawal is generally mild, making it difficult to model using acute mammalian models. To induce physical dependence in mammals typically requires administration of the benzodiazepine for days or weeks and withdrawal signs often do not develop until hours or even days following discontinuation of the drug (e.g., refs. 23-29). Using HPLC, we detected the presence of GABA (γ-aminobutyric acid) in planarians (manuscript submitted for publication) and speculated that planarians might display benzodiazepine-induced effects, possibly including withdrawal.

We used the usual metric of *p*LMV.[11] In the study,[30] planarians that were exposed to the benzodiazepines clorazepate or midazolam for 1 h and then tested in drug-free water displayed dose-related decreases in *p*LMV. The same effect was observed for zolpidem, which is a benzodiazepine receptor

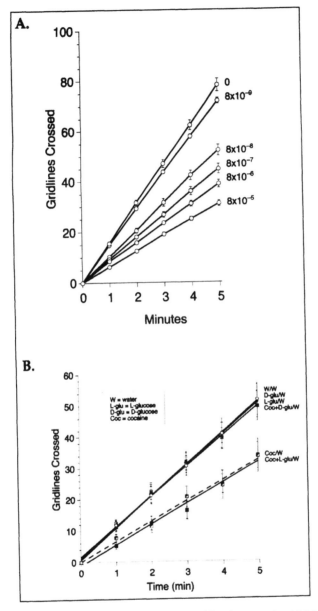

Figure 4. A) Abstinence-induced withdrawal (indicated by decrease in pLMV) following exposure to cocaine at the concentrations (M) indicated. B) pLMV (means ± S.D.) of planarians co-exposed to cocaine plus D- or L-glucose and then tested in water. Controls: planarians exposed to water, D- or L-glucose and then tested in water. N = 10 planarians per group. Reprinted with permission from references 10,21.

agonist, but is not chemically a benzodiazepine. In contrast, planarians that were pretreated for 1 hour with midazolam, clorazepate or zolpidem and then tested in the same concentration of the same drug, displayed no decrease in pLMV. Thus, the effect was only observed upon discontinuation (abstinence) of the drug (the results for midazolam are shown in Fig. 5A). Evidence of a direct pharmacologic effect was provided by the demonstration that clorazepate, midazolam and zolpidem

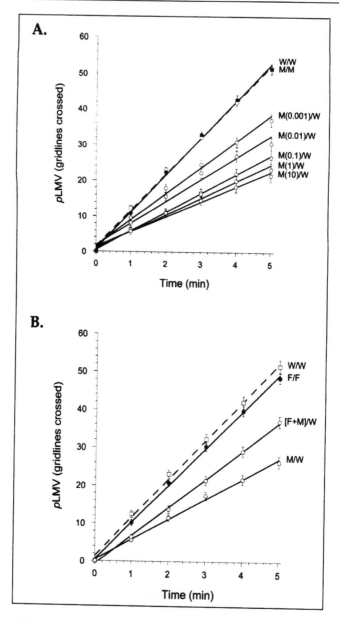

Figure 5. A) Withdrawal from midazolam in planarians as demonstrated by the dose-related decrease in ρLMV (expressed as mean ± s.e.m. cumulative number of gridlines crossed) following 1 h exposure to midazolam at the concentration (μM) indicated. B) Attenuation of the abstinence-induced withdrawal from midazolam in planarians by flumazenil, a benzodiazepine receptor antagonist. Reprinted with permission from reference 30.

withdrawal were all attenuated by the benzodiazepine receptor antagonist flumazenil (the results for attenuation of abstinence-induced withdrawal from midazolam are shown in Fig. 5B).

These results provide evidence that in the planarian model: benzodiazepine receptor agonists induce the rapid (≤1 h) development of physical dependence, independent of the chemical class of the agonist (i.e., benzodiazepine or other); there is rapid induction of abstinence-induced withdrawal (within minutes); and that the withdrawal appears to be mediated through a benzo-diazepine receptor antagonist-sensitive pathway.

Cannabinoid

In humans, a subset of chronic cannabis (marijuana) users exhibit a cluster of signs and symptoms after abrupt cessation of heavy use that includes irritability, sleep difficulties, restlessness, anxiety, depression, stomach pain and reduced appetite.[31-36] In animals such as rats, wet-dog shakes, pilo-erection, rearing, forelimb tremor, penile licking, mastication and scratching are noted.[37]

Planarians also display dose-related abstinence-induced withdrawal following exposure to the synthetic cannabinoid WIN 55212-2 ((R)-(+)-[2,3-Dihydro-5-methyl-3-[(4-morpholino)methyl] pyrrolo-[1,2,3-de]-1,4-benzoxazin-6-yl](1-naphthyl)methanone), but not its inactive enantiomer WIN 55212-3.[38] The WIN compound was used in these studies because Δ^9-THC, the active in-gredient of marijuana, is insoluble in aqueous solution and requires an ethanol-containing vehicle to be dissolved. WIN 55212-2 is soluble in a cremophor/water solution.[39]

Amphetamines

An amphetamine withdrawal syndrome has not been specifically identified in humans, but abrupt discontinuation following long-term excess amphetamine use can result in "extreme fatigue, mental depression and sleep pattern changes in EEG".[40] The behavior of methamphet-amine abusers undergoing the early stages of abstinence-induced withdrawal is characterized by inactivity, fatigue, anhedonia, anxiety and social inhibition.[41-43] Changes in brain dopamine levels have frequently been reported to occur during development of methamphetamine physical dependence, but other neurotransmitters and endogenous substances also appear to be involved (e.g., refs. 44-46). In animals, a type of behavioral dependence on methamphetamine use can be approximated using the place-preference model, but methamphetamine physical dependence and withdrawal is more difficult to model and quantify.[43,47,48] Thus despite the widespread abuse of amphetamines, the physiological and neurochemical mechanisms that underlie the development and expression of amphetamine physical dependence remain largely unknown.[43] One hurdle is the difficulty quantifying amphetamine withdrawal in mammals.[43,47,48] The planarian model offers a convenient alternative.

Amphetamine

Exposure to amphetamine slightly increases pLMA.[49] Thus amphetamine, a direct and indirect-acting dopamine and norepinephrine agonist, elicits the opposite effect on pLMV than does sulpiride, the dopaminergic receptor antagonist.[11] Evidence that dopamine is present in planarians, association of changes in planarian locomotor activity with dopamine receptors and that dopamine agonist-induced changes on motility are antagonized by the dopamine receptor antagonist haloperidol has been reported (e.g., refs. 5-8,14,50). The fact that amphetamine increases pLMV, albeit only modestly, is actually significant in terms of withdrawal studies with planarians, since decreased pLMV might be interpreted as resulting from some deleterious effect on planar-ian function (although multiple controls, including enantiomer-specific responding, etc. argue against this). The amphetamine results suggest that dopamine agonists and antagonists modulate the endogenous dopamine tone in planarians, i.e., agonists and antagonists increase and decrease, respectively, dopamine-mediated motility behavior.

Methamphetamine

Planarians display abstinence-induced withdrawal from methamphetamine (Fig. 6). Planarians exposed to methamphetamine and then tested in drug-free water display reduced pLMV compared to planarians exposed to water and tested in water and compared to methamphetamine-exposed planarians tested in the same concentration of methamphetamine. The magnitude of the withdrawal is concentration-related to the methamphetamine exposure.

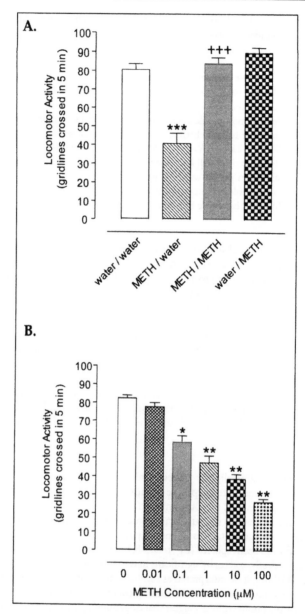

Figure 6. A) Withdrawal from methamphetamne (METH) in planarians. B) Withdrawal from methamphetamine (METH) is dose-related to methamphetamine exposure.

Conclusions

The planarian model of physical dependence/withdrawal does not comprise a substitute for more complex models in humans or other mammals. However, it does offer distinct advantages over some more complicated models, including: (i) a much simpler, yet sufficiently complex, nervous system—one that consists of many of the same neurotransmitter substances as in mammals, (ii) a well-defined withdrawal syndrome, (iii) a robust and sensitive metric that can be used to detect and quantify withdrawal—including withdrawal subtle and difficult to establish in mammals (e.g., benzodiazepines), (iv) fewer pharmacokinetic complications than mammals, (v) the oppportunity to study mechanism and 2nd-messenger pathways (see Chapter 7), (vi) ease of studying drug combinations (see Chapter 10) and (vii) convenience and flexibility.

References

1. Needleman HL. Tolerance and dependence in the planarian after continuous exposure to morphine. Nature 1967; 215:784-785.
2. Best JB, Rubinstein I. Maze learning and associated behavior in planaria. J Comp Physiol Psychol 1962; 55:560-566.
3. Lee RM. Conditioning of a free operant response in planaria. Science 1963; 139:1048-1049.
4. Griffard CD, Peirce JT. Conditioned discrimination in the planarian. Science 1964; 144:1472-1473.
5. Carolei A, Margotta V, Palladini G. Proposal of a new model with dopaminergic-cholinergic interactions for neuropharmacological investigations. Neuropsychobiology 1975; 1:355-364.
6. Algeri S, Carolei A, Ferretti P et al. Effects of dopaminergic agents on monoamine levels and motor behaviour in Planaria. Comp Biochem Physiol 1983; 74C:27-29.
7. Venturini G, Stocchi F, Margotta V et al. A pharmacological study of dopaminergic receptor in Planaria. Neuropharmacology 1989; 28:1377-1382.
8. Palladini G, Ruggieri S, Stocchi F et al. A pharmacological study of cocaine activity in Planaria. Comp Biochem Physiol C 1996; 115:41-45.
9. Raffa RB, Desai P. Description and quantification of cocaine withdrawal signs in Planaria. Brain Res 2005; 1032:200-202.
10. Raffa RB, Valdez JM. Cocaine withdrawal in Planaria. Eur J Pharmacol 2001; 430:143-145.
11. Raffa RB, Holland LJ, Schulingkamp RJ. Quantitative assessment of dopamine D2 antagonist activity using invertebrate (Planaria) locomotion as a functional endpoint. J Pharmacol Toxicol Meth 2001; 45:223-226.
12. Raffa RB, Stagliano GW, Umeda S. κ-Opioid withdrawal in Planaria. Neurosci Letts 2003; 349:139-142.
13. Pagliaro LA, Pagliaro AM. Comprehensive Guide to Drugs and Substances of Abuse. Washington DC: American Pharmacists Association, 2004; 8:
14. Passarelli F, Merante A, Pontieri FE et al. Opioid-dopamine interaction in planaria: A behavioral study. Comp Biochem Physiol C 1999; 124:51-55.
15. Tallarida RJ. Drug synergism and dose-effect data analysis. Boca Raton: Chapman and Hall/CRC Press, 2000; 150-151.
16. Arunlakshana O, Schild HO. Some quantitative use of drug antagonists. Brit J Pharmacol 1959; 14:48-58.
17. Raffa RB, Baron DA, Tallarida RJ. Schild (apparent pA₂) analysis of a κ-opioid antagonist in Planaria. Eur J Pharmacol 2006; 540:200-201.
18. Aceto MD, Bowman ER, Harris LS et al. Dependence studies of new compounds in the rhesus monkey, rat and mouse. NIDA Res Monogr 2002; 183:191-227.
19. Stevens CW. Opioid research in amphibians: a unique perspective on mechanisms of opioid analgesia and the evolution of opioid receptors. Revs Analgesia 2003; 7:69-82.
20. Pagliaro LA, Pagliaro AM. Comprehensive guide to drugs and substances of abuse. Washington DC: American Pharmacists Association, 2004; 81:
21. Umeda S, Stagliano GW, Raffa RB. Cocaine and κ-opioid withdrawal in Planaria blocked by D-, but not L-, glucose. Brain Res 2004; 1018:181-185.
22. Pagliaro LA, Pagliaro AM. Comprehensive guide to drugs and substances of abuse. Washington DC: American Pharmacists Association, 2004; 33:
23. Ryan GP, Boisse NR. Experimental induction of benzodiazepine tolerance and physical dependence. J Pharmacol Exp Ther 1983; 226:100-107.
24. McNicholas LF, Martin WR, Cherian S. Physical dependence on diazepam and lorazepam in the dog. J Pharmacol Exp Ther 1983; 226:783-789.
25. Rosenberg HC, Chiu TH. An antagonist-induced benzodiazepine abstinence syndrome. Eur J Pharmacol 1982; 81:153-157.
26. Pokk P, Zharkovsky A. Small platform stress attenuates the anxiogenic effect of diazepam withdrawal in the plus-maze test. Behav Brain Res 1998; 97:153-157.
27. Kaminski BJ, Sannerud CA, Weerts EM et al. Physical dependence in baboons chronically treated with low and high doses of diazepam. Behav Pharmacol 2003; 14:331-342.
28. Begg DP, Hallam KT, Norman TR. Attenuation of benzodiazepine withdrawal anxiety in the rat by serotonin antagonists. Behav Brain Res 2005; 161:286-290.
29. Listos J, Malec D, Fidecka S. Influence of adenosine receptor agonists on benzodiazepine withdrawal signs in mice. Eur J Pharmacol 2005; 523:71-78.

30. Raffa RB, Cavallo F, Capasso A. Flumazenil-sensitive dose-related physical dependence in planarians produced by two benzodiazepine and one nonbenzodiazepine benzodiazepine-receptor agonists. Eur J Pharmacol 2007; 564:88-93.

31. Cottler LB, Schuckit MA, Helzer JE et al. The DSM-IV field trial for substance use disorders: Major results. Drug Alcohol Depend 1995; 38:59-69.

33. Budney AJ, Radonovich KJ, Higgins ST et al. Adults seeking treatment for marijuana dependence: A comparison with cocaine dependent treatment seekers. Exp Clin Psychopharmacol 1998; 6:419-426.

32. Wiesbeck GA, Schuckit MA, Kalmijn JA et al. An evaluation of the history of a marijuana withdrawal syndrome in a large population. Addiction 1996; 91:1469-1478.

34. Haney M, Ward AS, Comer SD et al. Abstinence symptoms following oral THC administration to humans. Psychopharmacology 1999; 141:385-394.

35. Haney M, Ward AS, Comer SD et al. Abstinence symptoms following smoked marijuana in humans. Psychopharmacology 1999; 141:395-404.

36. Kouri EM, Pope HG. Abstinence symptoms during withdrawal from chronic marijuana use. Exp Clin Psychopharmacol 2000; 8:483-492.

37. Cui S-S, Bowen RC, Gu G-B et al. Prevention of cannabinoid withdrawal syndrome by lithium: Involvement of oxytoninergic neuronal activation. J Neurosci 2001; 21:9867-9876.

38. Rawls SM, Rodriguez T, Baron DA et al. A nitric oxide synthase inhibitor (L-NAME) attenuates abstinence-induced withdrawal from both cocaine and a cannabinoid agonist (WIN 55212-2) in Planaria. Brain Res 2006; 1099:82-87.

39. Rawls SM, Cabassa J, Geller EB et al. CB1 receptors in the preoptic anterior hypothalamus regulate WIN 55212-2 [(4,5-dihydro-2-methyl-4(4-morpholinylmethyl)-1-(1-naphthalenyl-carbonyl)-6H-pyrrolo[3,2,1ij] quinolin-6-one]-induced hypothermia. J Pharmacol Exp Ther 2002; 301:963-968.

40. Pagliaro LA, Pagliaro AM. Comprehensive guide to drugs and substances of abuse. Washington DC: American Pharmacists Association, 2004; 176:

41. Koob GF, Caine SB, Parsons L et al. Opponent process model and psychostimulant addiction. Pharmacol Biochem Behav 1997; 57:513-521.

42. Lago JA, Kosten TR. Stimulant withdrawal. Addiction 1994; 89:1477-1481.

43. McGregor C, Srisurapanont M, Jittiwutikarn J et al. The nature, time course and severity of methamphetamine withdrawal. Addiction 2005; 100:1320-1329.

44. Miyatake M, Narita M, Shibasaki M et al. Glutamatergic neurotransmission and protein kinase C play a role in neuron-glia communication during the development of methamphetamine-induced psychological dependence. Eur J Neurosci 2005; 22:1476-1488.

45. Rothman RB, Partilla JS, Baumann MH et al. Neurochemical neutralization of methamphetamine with high-affinity nonselective inhibitors of biogenic amine transporters: A pharmacological strategy for treating stimulant abuse. Synapse 2000; 35:222-227.

46. Rothman RB, Baumann MH. Balance between dopamine and serotonin release modulates behavioral effects of amphetamine-type drugs. Ann NY Acad Sci 2006; 1074:245-260.

47. Segal DS, Kuczenski R. Human methamphetamine pharmacokinetics simulated in the rat: Single daily intravenous administration reveals elements of sensitization and tolerance. Neuropsychopharmacology 2006; 31:941-955.

48. Segal DS, Kuczenski R, O'Neil ML et al. Prolonged exposure of rats to intravenous methamphetamine: behavioral and neurochemical characterization. Psychopharmacology (Berl) 2005; 180:501-512.

49. Raffa RB, Martley AF. Amphetamine-induced increase in planarian locomotor activity and block by UV light. Brain Res 2005; 1031:138-140.

50. Welsh JH, Williams LD. Monoamine containing neurons in Planaria. J Compar Neurol 1970; 138:103-116.

CHAPTER 10

Drug Combinations and Isoboles

Ronald J. Tallarida*

Abstract

This chapter discusses the quantitative pharmacology of drug combinations for drugs that produce overtly similar effects; more specifically we consider whether a combination of two agonists, or a combination of an agonist and an inactive second drug, produces a predictable effect or some exaggerated lesser or greater effect. The basis for answering this question (hence, predicting) employs the concept of dose equivalence and its consequence in producing and analyzing the Loewe isobologram. A sample of studies that have employed the isobologram are summarized here along with a more extended discussion of this usage for drug combinations that affect planarian locomotion.

Introduction

If two different agonist drugs produce overtly similar effects it is desirable to determine the magnitude of effect that occurs when they are administered together. The drug pair might be from any class such as analgesics, cardiac inotropic agents, cholesterol lowering agents, inhibitors of locomotion, anti-anxiety agents, etc. Their individual mechanisms of action might be quite different. Our concern is only that each has efficacy in producing a common effect that can be measured or that at least one of the two drugs has such efficacy. While it is reasonable to assume that the effect of any particular agonist dose combination can be predicted, the basis of such a prediction is not generally agreed upon (or understood) and therefore this topic has not yet found its way into our major textbooks of pharmacology, a fact that prompted this author's own monograph on drug combinations.[1] A common predictive assumption is that the effect of a dose combination is the simple sum of the effects produced by the drugs when each acts alone. This is much too simple and can be easily demonstrated to be incorrect because there is usually an upper limit to the effect of individual and combined drugs that is imposed by the test or by the physiological system. For example, antinociception, used as the effect metric in analgesic drug studies, will have some defined maximum (100% effect), perhaps based on a time (latency) or some other measure. If, for example, some dose of drug A yields an effect = 60% and a dose of drug B yields 70%, it is clearly meaningless to add these percent effects (130%?) when these doses are combined. Yet, the term "additive" is often used in this incorrect way in pharmacological testing and that usage prompts the question, If one does not add the effects, what is it that is added or should be added? That question and its answer are actually contained in the early works of Loewe[2,3] who introduced isobolographic analysis, a topic that is discussed next.

Drug Additivity and Isobolographic Analysis

Figure 1 illustrates the dose-effect curves for two different full agonists that produce the same effect and for which the relative potency (dose A/doseB) is a constant R. For a selected level of

*Ronald J. Tallarida—Department of Pharmacology, Temple University School of Medicine, 3420 N. Broad Street, Philadelphia, Pennsylvania, 19140 USA.
Email: ronald.tallarida@temple.edu

Planaria: A Model for Drug Action and Abuse, edited by Robert B. Raffa and Scott M. Rawls.
©2008 Landes Bioscience.

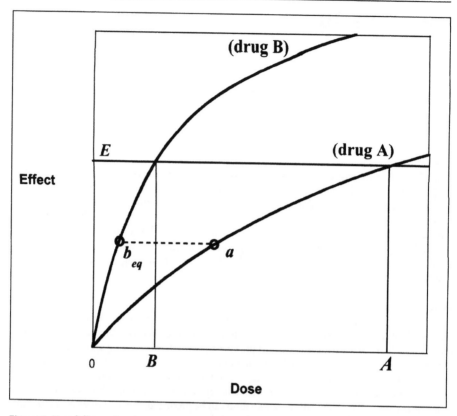

Figure 1. Two full agonists (maximum effects not shown) that have a constant relative potency R(= dose A/dose B) means that any dose a of drug A has its equivalent ($b_{eq} = a/R$) in terms of drug B (as shown by the broken line). For a specified effect E of interest, the individual agents require doses A and B as shown. In a combination containing dose a, some quantity b of drug B is needed so that the sum, $b + b_{eq}$ equals B. The quantity $(a + b)$ is therefore expected to give the effect of dose B when it acts alone. If this expectation is achieved we say that the drug interaction is "additive." The set of a, b pairs that give this effect constitutes the isobole of additivity as shown in Figure 2. In the additive case a plot of effect against total dose gives a curve that lies between the individual drug dose-effect curves.

effect (E) that is achieved by each alone, the needed dose is denoted A for drug A and B for drug B as shown. The constancy of R means that for any other dose pair (a,b) that produces a common effect, it is also true that $a/b = R$. In the figure that second effect level (shown as the broken line) is less than E. One can consider dose a to have its drug B-equivalent which is a/R and we denote that by b_{eq}. Thus, $b_{eq} = a/R$. We now consider all combinations (a,b) that give the effect E, an effect that required dose B when drug B acted alone. Toward that end, we use the above dose equivalence, adding b and b_{eq}, such that $b + beq = B$, which may also be written $b + a/R = B$. Division by B yields,

$$b/B + a/A = 1 \qquad\qquad (1)$$

The reasoning leading to equation (1) illustrates that what is added is dose b and the B-equivalent of dose a such that the sum is the dose B which yields effect E. This criterion, dose equivalence, is the basis used by Loewe for predicting how much of each is needed to achieve the effect level E. When equation (1) is viewed graphically it is seen to be a straight line (Fig. 2) in the first quadrant having intercepts A and B and the points on this line represent all combination dose pairs (a,b) that are expected to yield effect level E. The line segment is called the additive isobole for the

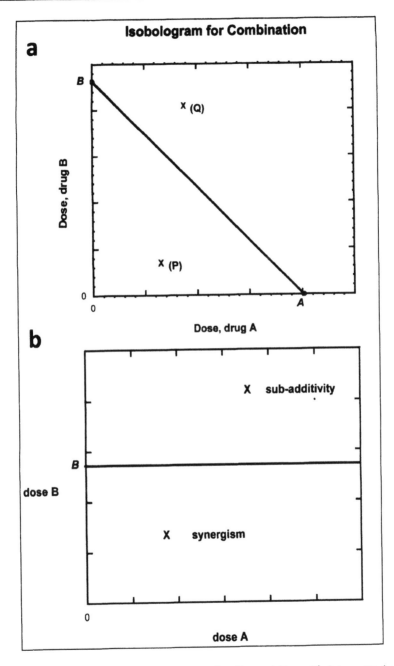

Figure 2. a) The additive isobole is shown as the diagonal line with intercepts *A* and *B*. Experimental dose pairs that give a point (Q) above the isobole are sub-additive, whereas those that plot below (P) are synergistic. b) A horizontal isobole applies when one agent (in this case drug A) lacks efficacy. The dose *B* indicates the potency of drug B and the addition of any quantity of drug A does not diminish the need for drug B to achieve the specified effect; hence, the isobole is horizontal. Experimental points above and below indicate sub-additive and synergistic interactions, respectively, as they do in the case when both drugs have efficacy.

specified effect. Equation (1) also may be derived by adding the drug A-equivalent of dose b to dose a, that is, $a + bR$ and equating this sum to A. Equation (1) is valid only if the relative potency R is constant at all effect levels. In other words, the equation follows from this assumption and the further assumption, implicit in the derivation, that the two drugs do not compete for a common binding site. Equation (1) does not define additivity; it is, instead, a consequence of independent action, a constant relative potency and the concept of dose equivalence. When experimental results yield points (dose combinations) that give the effect and which also lie on this line, we say that the combination is 'additive'; in alternative language, this case is said to be a case of zero interaction. In contrast, a lesser dose pair indicated by point P in Figure 2 may give the effect or a greater dose pair (point Q) may be required. The former indicates superadditivity (synergism) whereas the latter means sub-additivity. It will be shown subsequently that additivity may also produce curved isoboles. For the present, however, we will continue this discussion with the linear isobole of equation (1) and discuss its use in testing combinations.

Experiments with Drug Combinations

When drug combinations are tested and subjected to isobolographic analysis the aim is to determine the dose combinations that produce the specified effect magnitude from which the isobole was derived. If the tested combination (dose pair) for that effect plots as a point significantly below the isobole of additivity the combination is super-additive or synergistic. (Statistical aspects are given in several of this author's works; see, for example, refs 1, 4-6). In contrast, an experimental dose pair that plots above this line is sub-additive as shown in Figure 2. Experiments with numerous combinations have shown departures from additivity by employing isobolographic analysis. Drug combinations producing many different effects and novel experimental designs have been conducted. In most cases these used measures of an effect at a single time point. A notable early example is given by Gessner and Cabana[7] who described the hypnotic effect of chloral hydrate and alcohol using the righting reflex as the effect measure. Combinations of cocaine and buprenorphine were studied by Kimmel et al,[8] using locomotion as the endpoint. Pasternak's group examined *mu* opioids in combination as described in Bolan et al[9] and Field et al[10] studied combinations of gabapentin and NK1 antagonists using a model of neuropathic pain. Neostigmine was shown to interact synergistically with nonsteroidal anti-inflammatory drugs by Miranda et al[11] which prompted additional theoretical commentary by Tallarida.[12] Especially interesting are cases in which one of the two compounds lacks efficacy (a topic we will illustrate in a subsequent figure) but its presence enhances the effect of the active compound. An example of that situation was demonstrated in studies by Horan et al[13] in Porreca's laboratory that examined opioid delta receptor agonists with morphine. Tallarida et al[14] showed that glucosamine, which lacks efficacy in the mouse abdominal constriction test, significantly enhanced the antinociceptive activity of both ibuprofen and ketoprofen . This situation in which one drug lacks efficacy produces an additive isobole that is a horizontal line as seen in Figure 2 (lower). Numerous other studies have proceeded to analyze combinations with isobolograms. A very novel application, actually an extension of isobolographic analysis, is that in which a single drug is given at two different sites as described by Raffa et al,[15] who demonstrated site-site synergism. The basis for this application is also the concept of dose equivalence, i.e., the potency at one site has its equivalent value at the other site. Synergistic interactions have also been examined for enantiomers of an active compound, viz., tramadol, as shown by Raffa et al[16] That study showed that the (+) and (−) enantiomers of tramadol each independently produced centrally mediated antinociception in a standard test of antinociception in mice. The racemic compound was found to be more potent than the additive potency predicted from the enantiomers.

The several works cited here represent a small sample of the many studies that have employed isobolographic methods and, in most of these applications a linear isobole of additivity was the basis of the analysis for assessing sub-and superadditive interactions. This linear isobole was employed because the drug pairs showed a constant, or approximately constant, relative potency and, thus, the straight line isobole was the basis for the conclusions reached. As we will show below, certain experiments with planarians required a more generalized kind of isobole.

Drug Combinations in Planaria

Cocaine and WIN 55212-2

Raffa et al[17] studied the locomotive activity of planarians using the reduction in velocity (pLMV) as the effect in response to graded doses of WIN 55212-2, cocaine and combinations of the two agents in a fixed ratio mixture. The needed measurements were achieved by placemnt of the planarians into a clear plastic petri dish containing room-temperature tap water. Graph paper with gridlines uniformly spaced under the Petri dish allowed a measurement of velocity (pLMV) by counting the number of gridlines crossed or recrossed over a timed observation period. Prior to the measurement of velocity each planarian was placed into an individual 0.5 ml vial for 1 hour containing one of the following treatments: water, cocaine, WIN 55212-2 (each in several doses), or as fixed-ratio mixtures, also in various doses. The percent reduction in pLMV was taken as the effect in the determination of the dose-effect relations and consequent construction of dose effect curves. The curves for the individual agents (each alone) are shown in Figure 3A. It is seen that these curves have different slopes and different maxima and, thus, the relative potency of the drugs is not constant. In this situation, examined and analyzed by Grabovsky and Tallarida,[18] the application of the concept of dose equivalence still yields an isobole of additivity, but that isobole is not linear as shown in Figure 3B for the effect level = 50% of the maximum. Several different fixed ratio combinations of the two agents were studied and for each of these combinations the dose pair giving the 50% effect was determined (by linear regression of effect on log dose). These experimentally derived pairs are labeled A-E in the figure. The combinations denoted A and B are significantly below the additive curve indicating super-additive interactions. The other fixed-ratio combination values could not be distinguished from simple additivity. Point A, representing a fixed-ratio combination containing 23.5% WIN 55212-2 and 76.5% cocaine, has the ordinate value (cocaine component) = 0.00032 ± 0.00018 and point B, containing 12.4% WIN 55212-2 and 87.6% cocaine has the ordinate 0.00166 ± 0.00067. The additive ordinate corresponding to both points A and B (i.e., the value on the isobole) is 0.0103 ± 0.0014. Both experimental points are significantly below their corresponding additive values, thereby indicating a synergistic interaction. While the combinations shown as points C, D and E are also below the curve of additivity, these were not significantly different from values on the additivity curve.

Cocaine and U-50,488H

In another study with planarians by Raffa et al,[19] these investigators exposed the worms to cocaine, U-50,488H and combinations of the two agents for one hour. They then placed them in drug-free water. This procedure also displayed a dose-related reduction in pLMV. In contrast to the dose-effect curves of cocaine and WIN 55212-2, the dose effect relations for cocaine and U-50,488H (not shown) exhibited a relatively constant relative potency and, thus the additive isobole was a straight line (Fig. 4). Several fixed ratio combinations of these agents were tested, viz., 19:1, 1:19, 3:1, 1:1 and 1:3 (cocaine:U-50,488H). The 19:1 and 1:19 cocaine:U-50,488H ratios yielded D_{50} values whose plotted points on the isobologram were not significantly different from the expected additive values on the line, thereby indicating that there was no interaction between the two compounds at these ratios. However, the 3:1, 1:1 and 1:3 cocaine:U-50,488H ratios yielded D_{50} values that were each significantly different from the expected additive values for these ratios These results indicate that there is an interaction between the two compounds on withdrawal at these dose ratios and that the withdrawal (effect) of these combinations is less than additive, a situation usually termed "sub-additive."

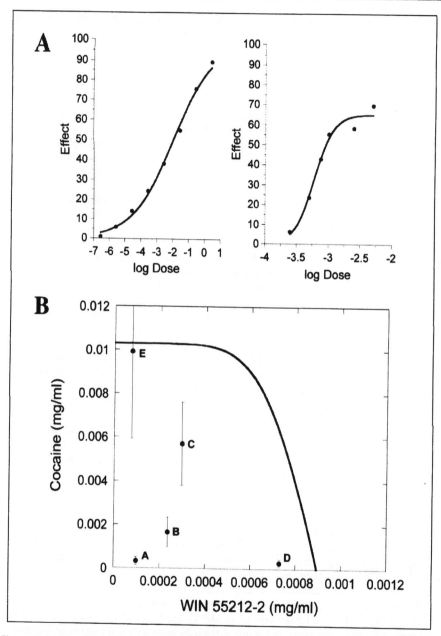

Figure 3. A) The dose-effect data for WIN 55212-2 (right) and for cocaine (left) were fitted to the smooth curves shown. These curves display D_{50} values for cocaine = 0.0102 mg/ml and for WIN 55212-2 the value 0.00089 mg/ml and these are shown as the intercept values in the nonlinear additive isobole curve shown in lower figure. B) This graph shows the additive isobole for the 50% effect (smooth curve) and the plotted points (dose pairs) for several tested combinations of cocaine and WIN 55212-2 on pLMV (the five points labeled A-E). The combinations denoted A and B are significantly below the additive curve coordinate values, thereby indicating super-additive interactions. The other experimental points, though below the curve, could not be distinguished statistically from the additive isobole. Reprinted with permission from reference 17..

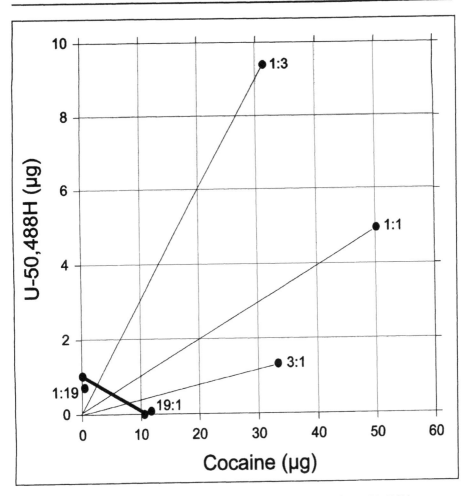

Figure 4. Planarians that were exposed to combinations of cocaine plus U-50,488H were ana-lyzed with the linear isobole of additivity shown as the short diagonal line. Linearity applied here because the individual agents displayed a constant relative potency. Three combination points (above the isobole), 1:3, 1:1 and 3:1(cocaine:U-50,488H) indicate sub-additive interac-tions of the two agents. The 19:1 and 1:19 cocaine:U-50,488H ratios yielded points that were simply additive in producing the 50% effect. Reprinted with permission from reference 19.

Conclusions

Combinations of active drugs are quite common in clinical situations and in recent years a large number of preclinical studies with combinations have been undertaken. The Loewe ap-proach, using the isobologram, has proved to be quite valuable in assessing interactions that are not expected from the individual drug potencies. The locomotion of planarians summarized here provides a rather precise metric for assessing the effects of certain drugs and drug combinations. Thus, this model of drug action is well suited to the kind of quantitation needed in drug studies. More generally, preclinical studies with drug combinations are enlightening and can serve as guides for ultimate clinical development and usage . A notable example arose from preclinical investiga-tions of antinociception with tramadol, acetaminophen and certain fixed ratio combinations of these by Raffa and Vaught that led to a U.S. patent (#5,336,691) and a new combination drug product. Also, determining the nature of a drug interaction is an important first step in exploring mechanism. The isobole is, however, still a bit mysterious to many investigators largely because of misunderstanding regarding the term "additive." As mentioned above this usage does not generally mean effect addition, although it can in some cases. Instead, additivity means that the dose of drug A can be converted to an equivalent dose of the other (drug B) and that equivalent plus the actual dose of B give a total dose B whose effect is predictable from drug B's dose-effect relation. When that concept is applied to drugs with a constant relative potency the analysis is simple because the resulting isoboles (for any chosen effect) are linear. However, when the potency ratio varies with the effect level, as is the case with planarian locomotive changes due cocaine and WIN 55212-2, the isobole of additivity is curved, thereby presenting a different kind of graph for distinguishing additive and non-additive interactions. The curved isobole also serves as a vivid reminder that drug additivity is not defined by equation (1) which is linear because that equation is applicable only when the relative potency is constant.

References
1. Tallarida RJ. Drug Synergism and Dose-Effect Data Analysis. In: Tallarida RJ, ed. Boca Raton: Chapman Hall, CRC Press, 2000.
2. Loewe S. Die Mischiarnei. Klin Wochenschr 1927; 6:1077-1085.
3. Loewe S. The problem of synergism and antagonism of combined drugs. Arzneimittelforschung 1953; 3:285-290.
4. Tallarida RJ, Cowan A, Adler MW. pA2 and receptor differentiation: A statistical analysis of competitive Antagonism. Mini-review Life Sci 1979; 25:637-654.
5. Tallarida RJ. An overview of drug combination analysis with isobolograms. Perspectives in pharmacology. J Pharmacol Exp Ther 2006; 319(1):1-7.
6. Tallarida RJ. Interactions between drugs and occupied receptors. Pharmacol Ther 2007; 113:197-209.
7. Gessner PK, Cabana BE. A study of the hypnotic effects and the toxic effects chloral hydrate and ethanol. J Pharmacol Exp Ther 1970; 174:247-259.
8. Kimmel HL, Tallarida RJ, Holtzman SG. Synergism between buprenorphine and cocaine on the rotational behavior of the nigrally-lesioned rat. Psychopharmacology 1997; 133:372-378.
9. Bolan EA, Tallarida RJ, Pasternak GW. Synergy between mu receptor subtypes: Evidence for functional interaction among mu receptor subtypes. J Pharmacol Exp Ther 2002; 303:557-562.
10. Field MK, Gonzalez MI, Tallarida RJ et al. Gabapentin and the NK1 antagonist CI-1021 act synergistically in two rat models of neuropathic pain. J Pharmacol Exp Ther 2002; 303:730-735.
11. Miranda HF, Sierraita F, Pinardi G. Neostigmine interactions with nonsteroidal anti-inflammatory drugs. Brit J Pharmacol 2002; 135:1591-1597.
12. Tallarida RJ, Miranda et al. Commentary on neostigmine interactions with non steroidal anti-inflammatory drugs. Brit J Pharmacol 2002; 135:1589-1590.
13. Horan P, Tallarida RJ, Haaseth RC et al. Antinociceptive interactions of opioid delta receptor agonists with morphine in mice: Supra- and subadditivity. Life Sci 1992; 50:1535-1541.
14. Tallarida RJ, Cowan A, Raffa RB. Antinociceptive synergy, additivity and subadditivity with combinations of oral glucosamine plus nonopioid analgesics in mice. J Pharmacol Exp Ther 2003; 307:699-704.
15. Raffa RB, Stone DJ Jr, Tallarida RJ. Discovery of 'self-synergistic' spinal/supraspinal antinoci-ception produced by acetaminophen (paracetamol). J Pharmacol Exp Ther 2000; 295:291-294.
16. Raffa RB, Friderichs E, Reimann W et al. Complementary and synergistic antinociceptive interaction between the enantiomers of tramadol. J Pharmacol Exp Ther 1993; 267:331-340.
17. Raffa RB, Stagliano GW, Tallarida RJ. Nonlinear isobologram and superadditive withdrawal from cocaine: cannabinoid combinations in planarians. Eur J Pharmacol 2007; 556:89–90.
18. Grabovsky Y, Tallarida RJ. Isobolographic analysis for combinations of a full and partial agonist: curved isoboles. J Pharmacol Exp Ther 2004; 310:981-986.
19. Raffa RB, Stagliano GW, Tallarida RJ. Subadditive withdrawal from cocaine κ opioid agonist combinations in planaria. Brain Res 2006; 1114:31-35.

CHAPTER 11

Analysis of Behavior in the Planarian Model

Cindy L. Nicolas, Charles I. Abramson and Michael Levin*

Abstract

Planaria species are powerful models for the study of drug effects and addiction on neural and cognitive function due to their tractability to cell-biological, pharmacological and molecular-genetic techniques. In order to fully capitalize on the many advantages of this system, it is necessary to be able to analyze behavior and learning in a quantitative manner in worms that have been treated with drug or RNAi reagents. Here, we give a brief overview of behavior and learning analysis in planaria. Classical data demonstrate that planaria can learn and exhibit many complex behaviors. We present a view of the next generation of work in this field; the development of automated, high-throughput platforms for analysis of planarian behavior will greatly enhance the integration of molecular genetics, nervous system structure and behavior in the same animal, extending our understanding of drug effects on cognition and opening the way for novel screening approaches to identify new compounds with important nootropic effects.

Introduction

This chapter presents a brief overview of research on learning in planarians, in order to illustrate the types of studies that can be performed in this model species to understand the effects of drugs on behavior and memory. We will provide some background on the history of planarian research and describe examples of planarian learning in the areas of habituation, classical conditioning and instrumental/operant conditioning. We would like to say at the outset that we believe that when proper procedures are followed, the data conclusively show that planarians can learn. Moreover, we believe that the study of planarian learning should be taken up once again because of the many advantages this animal has for the investigation of the biochemistry of learning and memory.

Planaria represent a critical breakthrough in the evolution of the animal body plan. It is the first organism to have both bilateral symmetry and encephalization, making it capable of detecting environmental stimuli quicker and more efficiently than the lower metazoans and therefore able to move in a directed fashion. They have developed sensory capabilities for the detection of light,[1,2] chemical gradients,[3,4] vibration,[5] electric fields,[6] magnetic fields[7,8] and weak γ radiation.[9] These reception mechanisms are integrated by the worm's nervous system into a rich and complex set of behaviors as they navigate their environment in the search for prey, mates, etc., and represent a fertile background of cognitive processes that may be modulated by drug exposure or withdrawal.

A key component of making use of this system for the study of drug effects and addiction is the ability to perform rigorous analyses of behavior and to establish paradigms in which true memory can be demonstrated. Thus, it is our goal to illustrate the types of analyses that have been done in planaria[10-14] and to sketch the outline of the next phase of this field of research, which centers

*Corresponding Author: Michael Levin—The Forsyth Institute and Department of Dev Biol, Harvard School of Dental Medicine, 140 The Fenway, Boston, MA 02115, USA. Email: mlevin@forsyth.org

Planaria: A Model for Drug Action and Abuse, edited by Robert B. Raffa and Scott M. Rawls. ©2008 Landes Bioscience.

around automated, quantitative characterization of behavior (and learning in particular). As the simplest animal with a bilaterally-symmetrical central nervous system comprised of neurons similar to our own,[15] planarians offer an opportunity to study the actions of chemicals in such neurons while working with an easily maintained organism that is highly tractable to pharmacological, surgical and molecular genetic manipulations.[16-19]

These flatworms also allow a researcher to investigate aspects of memory that are impossible to examine in other organisms, due to the planarians' robust ability to regenerate neural structures (see chapter by Oviedo and Levin in this volume). Many human neurotransmitters have been detected in extracts from planarians[20,21] and indeed it is clear that neurotransmitters have important functions in determining the structure of neural and nonneural components during planarian regeneration.[22] Over the last decade planarians have been used to model addiction and withdrawal for various psychoactive compounds such as cocaine,[23,24] though the effects such substances have on planarian learning and memory have generally been neglected.

Over the years there have appeared a number of excellent reviews on what has been called the "Planarian Controversy" and the interested reader is urged to consult these for the type of in-depth analysis that is not possible here.[25-27] Of special interest is a relatively recent review of Russian work.[28] There have also been a number of symposia in which a major focus was planarian learning.[29,30] For those interested in the planarian research pre-1940 should consult reference 31.

A Brief History of Learning Research in Planaria

One of the more exciting yet often overlooked chapters in the history of the experimental analysis of behavior is the work on learning and memory transfer in planarians.[32] This work captured the imagination of the public during the 1960s because of the possibility of directly transferring knowledge between organisms. The scientific literature of the mid 1950s through the early 1970s contained no less than 85 peer-reviewed papers designed to answer the question of whether planarians really learned and whether such learning could be transferred from one animal to another.

Controversies arose among laboratories that failed to replicate some findings, others which suggested alternative explanations and those which firmly held the belief that planarians learn. The interest in planarians by the general public was so great that the "Worm Runner's Digest" was created to aid amateur scientists who wanted to conduct their own studies. The Digest contained cartoons, tips and a few scientific papers. For those interested in more scholarly work on planarians the reader simply turned an issue upside down and found the "Journal of Biological Psychology." The Digest began publication in 1959 and the Journal was added to it in 1967. Both were published together until 1979.

The scientific study of learning in planarians was approached from two interrelated angles. The first was the comparative perspective in which the learning ability of planarians was compared to that of other animals in the expectation of illuminating the similarities and difference in learning across phyla.[31] Unfortunately the comparative analysis of planarian learning was not taken far enough and consisted mostly of a few isolated examples of punishment and maze learning. This lack of attention to training variables and conditioning phenomena would have serious consequences when planarians were subsequently used as a conditioning model.

In addition to the comparative aspect, there was interest in the use of planarians for what they can tell us about the biochemistry of learning and memory. This perspective—which led to what is now known as the simple-system strategy—attempts to uncover the neural circuits associated with a particular behavior.[33] It is ironic that contemporary discussions of the simple systems approach focus on various species of mollusk and insect[34] while neglecting work with planarians. As pointed out by McConnell and Shelby[27] the first experiments on memory transfer were conducted with planarians. However, there is no mention of these experiments (or of McConnell) in contemporary encyclopedic treatments as represented by either the Encyclopedia of Learning and Memory[35] or Comparative Psychology: A Handbook.[36]

The rationale behind the use of planarians as suitable material for simple system research was the work of Donald Hebb[37] and others who argued that learning produced physiological changes at the synapse. In 1953 Richard Thompson suggested to his fellow University of Texas graduate student, James McConnell, that since planarians are the first animal on the phylogenetic tree to possess the type of nervous system required by Hebb for learning, why not use planarians to test Hebb's theory?[38] The first study describing learning in planarians was published in 1955 using a classical conditioning paradigm in which light onset was paired with an electric shock.[39] The results suggested learning but unfortunately, because there was little comparative data, did not contain enough controls to rule out alternative explanations. McConnell followed this study with others that showed that classical conditioning can be transferred either by regeneration or by cannibalism.[40,41] Following publication of these studies, the "Planarian Controversy" began. The details of experiments that have been done in this field are crucial for those who seek to investigate the effects of drugs on memory and behavior in this species, because they hold many lessons that will enable the field to move forward and avoid the pitfalls that confounded some of the past attempts.

Learning

Learning has generally been classified into nonassociative and associative learning. Nonassociative learning is considered the more fundamental form, yet it shares many features with associative learning. The prototypical example of nonassociative learning is habituation. Associative learning is illustrated by such phenomena as classical conditioning, instrumental conditioning and operant conditioning. Information on how to conduct invertebrate learning experiments can be found in reference 42.

Habituation

Habituation is typically a decrease in some dependent variable (i.e., amplitude, probability) to a monotonously repeated stimulus. Stimuli that no longer transmit significant information tend to be ignored. Before a decrease in responsiveness can be considered a case of habituation several factors must be ruled out including effector fatigue, sensory adaptation, general experience with the training situation, temporal conditioning and the presence of chemical signals. Effector fatigue and sensory adaptation can often be ruled out by the addition of a distracter stimulus presented when the habituation criteria is met. If the response reappears to the original training stimulus following exposure to the distracter, an interpretation of the response decrease in terms of an inability to sense the training stimulus (sensory adaptation) or an inability to make a response even though the stimulus can still be sensed (effector fatigue) is unwarranted.

The number of planarian experiments on habituation is surprisingly quite small. This is unfortunate for a number of reasons. First, habituation shares many phenomena with associative learning. These include spontaneous recovery, stimulus intensity effects and generalization.[43] Second, habituation is a fine comparative tool through which species can easily be compared. Third, habituation can be used as a control for planarian transfer of training experiments that have often been criticized as representing the transfer of general excitatory tendencies rather than a specific behavior. If both excitatory and inhibitory responses can be transferred, the case for physiological and biochemical correlates of the transfer phenomena is greatly enhanced.[44,45]

Although not the primary object of investigation, habituation to light is often assessed prior to using light as a conditioned stimulus in classical conditioning experiments. Several studies have shown that planarians will decrease their responses to light following repeated presentations.[46] Whether such decreased responsiveness is the result of habituation or some other process is not known, because habituation controls, understandably, were not in place. Using habitat rotation and light Walter[47] observed a decrease in responsiveness to repeated stimulation but again controls were not in place.

Perhaps the best studies of habituation in planarians were performed by Westerman.[44,45] Westerman presented planarians with 3 seconds of light over the course of 25 trials per day for 20 days. The dependent variable was contraction and animals had to meet a criterion of 50 trials with no contraction to light. Animals were not only able to reach criteria but such habituation survived regeneration and could be retained for 7 weeks.

Classical Conditioning

To function successfully in a changing environment, planarians must not only learn new behaviors but also call on reflexive responses in new contexts. Classical conditioning is an example of associative learning in which behavior is altered by the pairing of stimuli, one of which is effective in eliciting a biologically important reflex. A common feature of classical conditioning is that the conditioned stimulus (CS) and unconditioned stimulus (US) are presented independently of the organism's actions.

Before it can be concluded that animals have learned to associate a CS and US pseudoconditioning, or "false" learning not based on the explicit pairing of the CS and US, must be ruled out. The easiest way to do this is to use a between group design where a group of animals receiving paired CS-US presentations is compared to another group receiving unpaired CS/US presentations. A statistically significant difference between paired and unpaired groups is evidence for learning. Generally, the intertrial interval for the unpaired group should be half the interval used in the paired group. In this way the time between the CS presentations (the intertrial interval) is approximately equal between experimental and control groups. If the same intertrial interval is used, the time between CS presentations in the unpaired group will be twice as long as in the paired group and any difference between the groups may not be attributed to learning.

In addition to using a group design, a good protocol calls for a within subject design in which two CSs are used—one of which is paired with the US. Evidence for learning is provided by a statistically significant difference between responses associated with the CS paired with the US and a second CS that is not. Data from both group and within subject designs together strongly supports conditioning.

The vast majority of planarian learning experiments were performed with classical conditioning paradigms using light or vibration as a conditioned stimulus and shock as the unconditioned stimulus. The original Thompson and McConnell[39] experiment served as the model with the technique being refined over the years.[26] Unfortunately an unpaired group receiving the same number of CSs and USs was not run in the original demonstration and this left the investigation open to criticism.[48]

Since 1955, a number of studies have employed suitable control groups and have demonstrated classical conditioning. These include backward, delayed, simultaneous conditioning and various types of pseudoconditioning controls.[49,50] Using light and vibration as conditioned stimuli, discrimination learning has also been demonstrated in which the planarian is trained to respond to a CS paired with shock and one that is not paired.[49,51-53] Moreover, conditioning has been demonstrated both in acquisition and in extinction.[54,55] Of 71 experiments investigating various aspects of classical conditioning in planarians 54 (76%) have produced positive results.[26]

Instrumental and Operant Conditioning

In instrumental and operant conditioning, a contingency is arranged between a motivationally significant stimulus and a specific behavior. The planarian learns that consequences occur as a result of its actions. Instrumental and operant conditioning are generally considered more complex than classical conditioning although they share many properties.

For instrumental and operant conditioning, controls must be implemented to ensure that the planarian is indeed learning the consequences of its actions. The problem of control is arguably not as great as in classical conditioning because in principle the instrumental and operant response should be a behavior that is not in the repertoire of control animals. For example, if several planarians having been taught to negotiate a maze are placed in a general population containing untrained planarians, it should be an easy matter to determine trained from untrained. In addition, many maze experiments have a built in control by requiring the animal to make a discrimination.

The vast majority of instrumental/operant studies using planarians have employed various types of mazes often using escape from shock or return to the home container as a reward. Several maze configurations have been used including standard T and Y forms. Hexagonal and multi-unit mazes have also been used.[26] Maze performance is often variable with early success giving way to instability and eventually a refusal to run the maze.[56] The most successful maze technique has been developed by Best[57] in which planarians are faced with three arms, any of which can be drained of water. Animals were able to form a light-dark discrimination by entering the correct arm and being reinforced with water. Even here, however, performance can become unstable. Positive results have been obtained with a hexagonal maze that reduced the intertrial interval and minimized handling.[58,59]

Operant situations have also been used with better success than that reported for mazes. Lee[60] developed a paradigm in which a planarian confined to a cylinder was trained to interrupt a photobeam that briefly terminated a bright light. This work has been replicated.[57,61]

One of the most successful instrumental training techniques for planarians is known as the "Van Oye Maze" in honor of its developer although in contemporary usage it does not resemble a maze.[62] In this situation food is suspended from a glass rod into a beaker containing several planarians. The rod can be placed to any desired depth with the worms crossing the surface of the water and down the rod to feed. As training progresses the rod is moved to a deeper depth. During the testing phase a clean rod is placed at the final training depth and the results show that significantly more trained planarians are at the end of the rod than untrained planarians. These results have been replicated several times and constitute some of the strongest evidence of the ability of planarians to learn by consequences.[25,62,63]

Past Drug Experiments

Planarians have been studied since at least the 18th century[64] and studies have been done with them on the effects of drug treatment and withdrawal.[23,24,65] Such studies include looking at the effects of caffeine on respiration rate[66] and the development of a dependence on morphine.[67] The most recent drug abuse-related studies have focused on commonly abused drugs such as cocaine and opioids. To evaluate the effects of drug treatments innate behaviors are typically observed. Evidence for drug impacts has included stereotypically disrupted movements such as "corkscrew" and "head bop" and "tail twist,"[24] a loss of the negative phototaxis planaria normally display[68] and hypokinesia.[69] Rarely have these studies considered the impacts of abuse-prone drugs on learning and memory, as did one methamphetamine study.[70] Displays of learning and memory are complex, requiring the correct coordination of central nervous system and peripheral nervous system activities. Therefore, discovering specific disruptions in these processes has the potential to shed considerable new light on the actions drugs have on the nervous system.

Where Are We Now?

As Corning and Kelly lamented in 1973, interest in the use of planarians for research on learning and memory has declined and such work today is rare. Planarians were the first invertebrate to be used in simple system research and to stimulate interest in the biochemistry of learning and memory. These animals are easy to maintain and easy to manipulate for learning and biochemical experiments. Both cut and uncut worms can be readily exposed to drugs and the increasingly popular use of RNAi in planarian work provides an opportunity to interrupt a particular biochemical pathway as well.[71,72] Modern molecular studies are beginning to address the cell biology and genetics of neural networks in planaria.[73]

Unfortunately, in our view, much time was wasted during the "Golden Age of Planarian Research" on experiments with problematic design investigating whether these animals could learn and whether learning or some non-associative effect was transferred either by regeneration or by cannibalism. Little effort was made toward advancing past basic demonstrations of conditioning phenomena. For example, questions regarding "cognitive" effects such as latent inhibition, US pre-exposure, second-order conditioning, blocking, overshadowing and within compound associations as well as the effects chemical compounds or RNA-mediated interruptions of endogenous pathways could have on such learning all have remained uninvestigated in planarians.

On the positive side there is little doubt that planarians can learn. Progress has also been made in identifying critical factors that influence such learning. These factors include the presence of slime, housing, diurnal cycles, medium associated with housing, training and testing, species differences, apparatus and such training variables as stimulus intensity, intertrial interval and number of stimulus pairings.[26] Biological research on stem cells and regeneration has created fresh interest in planarians as a model organism, leading to increased knowledge of planarian genetics and neuroanatomy. The species *Schmidtea mediterranea* has a genomic database online at http://smedgd.neuro.utah.edu/, which should aid researchers working with this species.[74,75] This type of newly available knowledge of planarian genomics suggests experiments that may combine studies of drug treatments with RNAi targeted at specific neural receptor proteins.

The Future of Studies in Planarian Behavior

We believe the time is right to re-invigorate planarian research on learning and memory utilizing the most powerful tools of genetics, cell biology and biochemistry. It is crucial however, to be able to perform integrative studies that can address the whole path leading from genetics to nervous system structure and ultimately to behavior and to assess results without subjectivity or experimenter variations. Thus, what is required now is the development of computerized, automated, high-throughput devices that can be used to efficiently and quantitatively characterize behavior of animals that have been manipulated pharmacologically or genetically (e.g., knockouts of specific neurotransmitter receptors). Moreover, functional analyses of the effects of drugs on memory and learning will require a robust assay that can produce a "strong" memory that persists through the course of a specific experiment and that are easy to measure. This assay also must be reproducible and efficient enough to accommodate large enough sample sizes to ensure high statistical significance of results despite variability among individual animals.

Most available systems for automated behavior analysis focus on rodents,[76-82] but a few systems have been developed for small species that are amenable to behavioral screening and large population studies such as crustaceans,[10,46,60,83-85] zebrafish,[10,46,60,83-85] and *C. elegans*.[86] Unfortunately, none of the existing (commercially-available) solutions meet the necessary criteria; most lacking is the ability to modulate the environment of the animal in real-time: that is, not only to analyze behavior but to provide feedback (positive and negative reinforcement) individually to each animal so that automated training and testing of recall can occur.

The ideal behavior analysis device would have the following properties. 1) Convenient programming of essential control conditions such as yoked controls, in which control animals receive rewards and punishments based on the behavior of other animals being trained. This controls for the effects of the rewards/punishments per se, when these are not linked to the actual behavior of the animals. 2) Full automation to exclude experimenter effects and subjective scoring, enable rich data acquisition and ensure reproducibility. 3) Programming of individualized combinations of environmental cues and feedback (reward/punishment) in real time to accommodate different rates of memory and learning among animals within a sample group. 4) Flexibility in designing and programming training and testing paradigms, allowing the whole parameter space to be explored for the important variables in a learning trial. 5) Accommodation of large sample sizes for statistical and screening purposes (high-throughput), allowing many different treatments to be efficiently analyzed. 6) Recording of all primary data, for subsequent review, analysis and transfer to other laboratories. This is essential to allow other investigators to mine the data in different ways and could be used over the web as a training opportunity for students to interact with such a device remotely. 7) Efficient modular design to enable deployment in other laboratories; this is very important since a considerable degree of discord during the early days of planarian research was due to the difficulty of reproducing training environments precisely in different labs. A fully-specified automated system that anyone could buy would enable investigators to replicate experimental designs perfectly and also avoid experimenter effects (providing completely objective scoring). Of course, such devices would also be useful for other model species such as zebrafish which also offer advantages of genetic or optical tractability.

Use of computer controlled training systems would contribute great clarity to this field, by eliminating potential variation between experimenters as well as allowing many animals to be run in parallel. It would also reduce artifacts due to handling and provide for much greater statistical power. A solid foundation, in terms of modern training techniques and proper controls can help push through the controversies of the past and allow this simple animal to shed light on the fundamental basis for learning and memory. A proposed design, which is currently guiding the efforts in our group, is shown in Figure 1. A rack-mounted set of drawers contains pull-out trays of 5 × 5 cells. Each such cell is a Petri dish containing a photoelement (for tracking worm behavior), light-emitting diodes (to establish light and dark quadrants as stimuli) and electrodes (to produce shock for negative reinforcement). This is in effect a set of Skinner chambers, where each animal can be independently trained on an almost limitless number of tasks (e.g., "stay in center", "follow the lit up quadrant", "keep moving", "move when a light flashes", etc.). A centralized processor controls the environment in each dish based on the behavior of the given worm.

The algorithm for such a system is schematized in Figure 2A-F. Once the user sets up the training parameters for all of the dishes, there is a repeated cycle (providing a consistent environment for

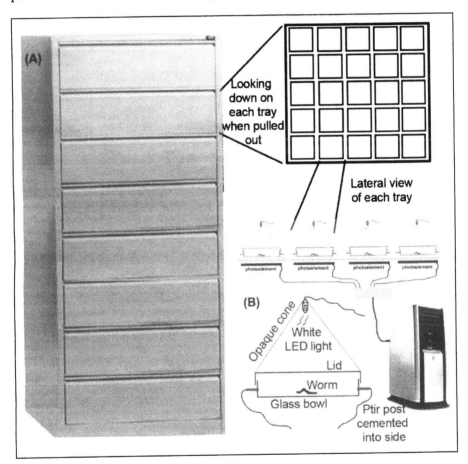

Figure 1. Schematic of high-throughput device for analysis of planarian behavior. A) Rack-mounted system containing pull-out trays of individual dish arrays. Each dish consists of a disposable Petri dish, a photoelement beneath and a cover that has an array of light-emitting diodes (LEDs) that light up various areas of the dish without introducing heating and an insert with inert electrodes for providing shock. The whole system is illuminated by a weak light in the far red end of the spectrum which is invisible to the worms but allows the camera to capture their movements. B) Close-up of one such dish. The LEDs and electrodes are connected to a custom programmable logic array that transduces data and effector commands to and from the software (running on a computer).

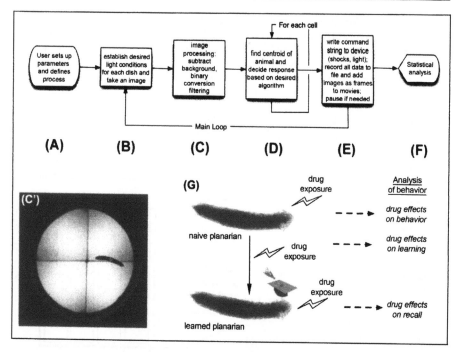

Figure 2. Schematic of algorithm for automated analysis of worm behavior. A) The user sets up the trial, indicating the light conditions, what behaviors will be rewarded and punished for each dish and the parameters of trial length, rest periods, shock strength, light brightness, etc. Throughout the trial, the device will repeat a cycle consisting of (B) changing the lights if needed and grabbing images of each dish, (C) low-level image processing, (D) detecting the centroid of the worm's position (actual such image shown in C', centroid indicated by asterisk), (E) sending commands to shock those worms that are to be punished and recording all the data to files on disk. The cycle can be performed at any rate (e.g., 5 complete cycles per second) and (F) statistical analysis is performed when it is completed. G) Schematics of how this could be used to investigate drug effects or withdrawal. The device will be used to characterize the behavior (collecting data on rate of movement, preferences for the edge vs bottom of dish, aversion to light, sensitivity to shock, etc.,) of worms exposed to various pharmacological reagents or removed from exposure to drugs to which they are addicted (vs controls). This can be done in the absence of learning (for basic effects on behavior), during training (to characterize effects on learning) and during recall (to study effects on duration or quality of recall).

some number of hours or days). This cycle consists of image processing to detect the position of each worm (Fig. 2C), decisions (determined by the user-specified learning task) as to whether each worm is to be punished or rewarded and a corresponding change in the light and shock environment of the dish. All of the data (coordinates of the worms' movements plus real-time movies of their activities) are recorded to disk and may be made instantly available via the Internet. In simpler trials (with no rewards or punishments), the system can simply gather quantitative data on the baseline behavior of control or drug-treated worms and their response to light, shock and vibration.

Such a paradigm would be a powerful addition to studies on drug effects (Fig. 2G). For example, drugs can be added without learning (to examine effects on behavior), during training (to examine effects on the magnitude and time-course of learning), or during testing (to examine effects on length and quality of recall). Naturally, drugs can be withdrawn from addicted animals to be tested in a similar fashion.

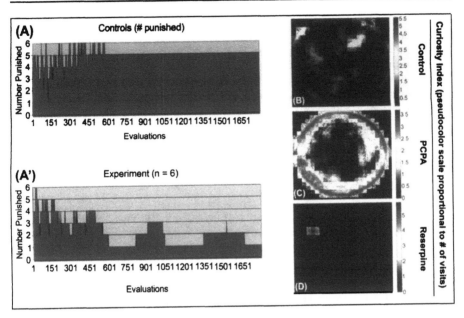

Figure 3. Sample data from prototype behavior analysis device. One of the output streams is shown here for a typical experiment, where worms are trained against preference (punished by a weak electric shock for avoiding a brightly lit region in the dish). Ignoring the many metrics that are also gathered by this system (average velocity, % of time at the edge vs center of dish, etc.), this panel focuses on the progress of a learning trial. Yoked controls (worms that are punished according to the behavior of the trained worms and thus have no opportunity to discern a cause-effect relationship), show no improvement over time (A). In contrast, the trained group shows a considerable decrease in their presence in the normally-preferred dark portion of the dish (A'). Effects of drug exposure and withdrawal can easily be detected by such metrics when applied during training or testing phases. Effects of drugs on motor behavior can also be studied via a "curiosity plot", where the device pseudocolors the area of a dish based on how often the worm visited that region. This allows one to determine, at a glance (and without recourse to the timelapse movies), how much of a dish was explored by a given animal. B) A control animal that explored the dish and settled in the upper-right-hand corner. C) A worm treated with PCPA, which exhibited the expected frantic activity of continually circling around the edge of the dish. D) A worm treated with reserpine (targeting serotonin levels), which exhibited the expected lethargy and never moved from its spot. These data and relevant methods/controls are given in reference 87.

Seeking to develop a prototype for this next step in the field, we produced a small system (suitable for 12 planaria at a time) and illustrated its use in the automated training and testing of control and drug-treated worms.[87] The reduced handling, objective scoring of behavior and quantitative real-time analysis resulted in a very powerful system that can easily be applied to the study of drug effects; an example application is shown in Figure 3. The statistical strength of numbers and the reduced tedium (compared to manual training) opens the way for analyses to be made on the behavioral effects from large numbers of psychoactive chemicals and genetic constructs on behavior, learning and memory. Combinations of chemicals can be observed as can the effectiveness of many different treatments on animals undergoing withdrawal from addictive drugs. With proper controls and careful experimental design, automation that eliminates the possibility of experimenter biases not only offers an opportunity to more clearly illuminate the controversies of the past, but also provids a solid foundation for screening potential psychoactive drugs and treatments in vivo in a simple lower organism. Moreover, if expanded to tens of thousands of cells in a much larger device, this system offers the possibility of screening drug and RNAi libraries for nootropic compounds (e.g., increases of learning rate), something that is not possible using today's screening platforms that are based on cells in culture or yeast.

As Sarnat and Netsky (1985) noted, the planarian brain is rather unique. The ratio of the brain to body weight is similar to that of the rat. Planarian neurons contain serotonin, acetylcholine, norepinephrine and epinephrine and there are many morphological and electrophysiological features analogous to the vertebrate brain. As the obstacles that made planarian behavioral research problematic in past decades are overcome and the work is integrated with the tremendously powerful tools of modern molecular genetics, a new era of planarian research is beginning. Planarians are deceptively simple organisms that utilize an organized central nervous system and some of the same neurochemicals as higher vertebrates. They clearly have much yet to tell us about the biochemistry of learning and memory and about the ways in which drug compounds impact cognition.

Acknowledgements
We thank Emily Yuan, Caitlin Mueller, Caitlin Hicks and Debbie Sorocco for early work on behavioral analysis of planaria in our lab. This work was supported by NSF grant IBN #0347295 and NIH grant HD055850-01. It was written in a Forsyth Institute facility renovated with support from Research Facilities Improvement Grant Number CO6RR11244 from the National Center for Research Resources, National Institutes of Health.

References
1. Brown HM, Ito H, Ogden TE. Spectral sensitivity of the planarian ocellus. J Gen Physiol 1968; 51(2):255-260.
2. Brown F, Park Y. A persistent monthly variation in responses of planarians to light and its annual modulation. Internat J Chronobiol 1975; 3(1):57-62.
3. Miyamoto S, Shimozawa A. Chemotaxis in the freshwater planarian Dugesia-japonica-japonica. Zoological Science (Tokyo) 1985; 2(3):389-396.
4. Mason PR. Chemo-klino-kinesis in planarian food location. Anim Behavr 1975; 23(2):460-469.
5. Fulgheri D, Messeri P. The use of 2 different reinforcements in light darkness discrimination in planaria. Bollettino—Societa Italiana Biologia Sperimentale 1973; 49(20):1141-1145.
6. Brown HM, Ogden TE. The electrical response of the planarian ocellus. J Gen Physiol 1968; 51(2):237-253.
7. Brown FA. Effects and after-effects on planarians of reversals of the horizontal magnetic vector. Nature 1966; 209(22):533-535.
8. Brown F, Chow C. Differentiation between clockwise and counterclockwise magnetic rotation by the planarian dugesia-dorotocephala. Physiol Zool 1975; 48(2):168-176.
9. Brown F, Park Y. Seasonal variations in sign and strength of gamma-taxis in planarians. Nature 1964; 202:469-471.
10. Tiras KP, Aslanidi KB. Device for the graphic recording of planaria behavior. Zhurnal Vysshei Nervnoi Deyatelnosti Imeni I P Pavlova 1981; 31(4):874-877.
11. Riccio D, Corning WC. Slime and planarian behavior. Psychological Record 1969; 19(3):507- and.
12. Reynierse JH. Reactions to light in four species of planaria. J Comp Physiol Psychol 1967; 63(2):366-368.
13. Pearl R. The movements and reactions of fresh-water planarians: a study in animal behavior. Q J Microsc Sci 1903; 46(v):509-714.
14. Walter H. Reactions of planarians to light. J Exp Zool 1907; 5(1):37-162.
15. Sarnat HB, Netsky MG. The brain of the planarian as the ancestor of the human brain. Can J Neurol Sci 1985;
16. Eisenstein EM. Selecting a model system for neurobiological studies of learning and memory. Behav Brain Res 1997; 82(2):121-132.
17. Sanchez Alvarado A, Reddien PW, Bermange A et al. Functional Studies of Regeneration in the Planarian Schmidtea mediterranea. Dev Biol 2003; 259:525.
18. Reddien PW, Sanchez Alvarado A. Fundamentals of planarian regeneration. Annu Rev Cell Dev Biol 2004; 20:725-757.
19. Newmark PA. Opening a new can of worms: a large-scale RNAi screen in planarians. Dev Cell 2005; 8(5):623-624.
20. Ribeiro P, El-Shehabi F, Patocka N. Classical transmitters and their receptors in flatworms. Parasitology 2005; 131(Suppl):S19-40.

21. Welsh JH, Williams LD. Monoamine-containing neurons in planaria. J Comp Neurol 1970; 138(1):103-115.
22. Villar D, Schaeffer DJ. Morphogenetic action of neurotransmitters on regenerating planarians—A review. Biomed Environ Sci 1993; 6(4):327-347.
23. Raffa RB, Valdez JM. Cocaine withdrawal in planaria. Eur J Pharmacol 2001; 430(1):143-145.
24. Raffa RB, Desai P. Description and quantification of cocaine withdrawal signs in planaria. Brain Res 2005; 1032(1-2):200-202.
25. Corning WC, Riccio D. The planarian controversy. In: Byrne W, ed. Molecular Approaches to Learning and Memory. New York: Academic Press, 1970:107-150.
26. Corning WC, Kelly S. Platyhelminthes: the turbellarians. In: Corning WC, Dyal JA, Willows OD, eds. Invertebrate Learning. New York: Plenum Press, 1973; I:
27. McConnell JV, Shelby JM. Memory transfer experiments in invertebrates. In: Ungar G, ed. Molecular Mechanisms in Memory and Learning. New York: Plenum Press, 1970:71-101.
28. Sheiman IM, Tiras KL. Memory and morphogenesis in planaria and beetle. In: Abramson CI ZPS, Burmistrov YM, eds. Russian Contributions to Invertebrate Behavior. Westport: Praeger, 1996:43-76.
29. Byrne WL. Molecular Approaches to Learning and Memory. New York: Academic Press, 1970.
30. Corning WC, Ratner SC, American institute of biological sciences. Chemistry of Learning; Invertebrate Research. New York: Plenum Press, 1967:
31. Warden CJ, Jenkins TN, Warner LH. Comparative Psychology. New York: Ronald Press, 1940:2.
32. Rilling M. The mystery of the vanished citations: James McConnell's forgotten 1960s quest for planarian learning, a biochemical engram and celebrity. Am Psychol 1996; 51(6):589-598.
33. Carew TJ, Sahley CL. Invertebrate learning and memory: From behavior to molecules. Annu Rev Neurosci 1986; 9:435-487.
34. Kandel ER, Schwartz JH, Jessell TM. Principles of Neural Science. 4th ed. New York: McGraw-Hill Health Professions Division, 2000.
35. Byrne JH. Encyclopedia of Learning and Memory. Farmington Hills: Thomson/Gale, 2003;
36. Greenberg G, Haraway MM. Comparative psychology: A Handbook. New York: Garland Pub., 1998:
37. Hebb DO. A Textbook of Psychology. Philadelphia: Saunders, 1958;
38. McConnell JV. The modern search for the engram. In: McConnell JV, ed. A Manual of Psychological Experimentation on Planarians. 2nd ed. Ann Arbour: Journal of Biological Psychology, 1967.
39. Thompson R, McConnell JV. Classical conditioning in the planarian Dugesia dorotocephala. J Comp Physiol Psychol 1955; 48:65-68.
40. McConnell JV. Memory transfer through cannibalism in planarians. J Neuropsychiatr 1962; 3:42-48.
41. McConnell JV, Jacobson AL, Kimble DP. The effects of regeneration upon retention of a conditioned response in the planarian. J Comp Physiol Psychol 1959; 52:1-5.
42. Abramson CI. A Primer of Invertebrate Learning : The Behavioral Perspective. 1st ed. Washington DC: American Psychological Association, 1994;
43. Thompson RF, Spencer WA. Habituation—A model phenomenon for study of neuronal substrates of behavior. Psychol Rev 1966; 73(1):16.
44. Westerman RA. Somatic Inheritance of habituation of responses to light in planarians. Science 1963; 140(3567):676-677.
45. Westerman RA. A study of habituation of responses to light in the planarian Dugesia dorotocephala. Worm Runner's Digest 1963; 5:6-11.
46. Corning WC, Freed S. Planarian behaviour and biochemistry. Nature 1968; 219(160):1227-1229.
47. Walter HE. Reactions of planarians to light. J Exp Zool 1908; 5:35-163.
48. Jensen DD. Paramecia, planaria and pseudolearning. Anim Behav 1965; (Suppl 1):9-20.
49. Jacobson A, Horowitz S, Fried C. Classical conditioning, pseudoconditioning, or sensitization in the planarian. J Comp Physiol Psychol 1967; 64(1):73-79.
50. Vattano FJ, Hullett JH. Learning in planarians as a function of interstimulus interval. Psychon Sci 1964; 1:331-332.
51. Block RA, McConnell JV. Classically conditioned discrimination in the planarian, Dugesia dorotocephala. Nature 1967; 215(109):1465-1466.
52. Fantl S, Nevin JA. Classical discrimination in planarians. Worm Runner's Digest 1965; 7:32-34.
53. Griffard CD, Peirce JT. Conditioned discrimination in the planarian. Science 1964; 144:1472-1473.
54. Yaremko RM, Kimmel HD. Two procedures for studying partial reinforcement effects in classical conditioning of the planarian. Anim Behav 1969; 17(1):40-42.
55. Crawford T, Livingston P, King F. Distribution of practice in the classical conditioning of planarians. Psychon Sci 1966; 4:29-30.
56. Corning WC. Evidence of right-left discrimination in planarians. J Psychol 1964; 58:131-139.
57. Best JB. Behaviour of planaria in instrumental learning paradigms. Anim Behavr Supp 1965; 13(Suppl 1):69-75.

58. Roe K. In search of the locus of learning in planarians. Worm Runner's Digest 1963; 5:16-24.
59. Humpheries B, McConnell JV. Factors affecting maze learning in planarians. Worm Runner's Digest 1964; 6:52-59.
60. Lee RM. Conditioning of a free operant response in planaria. Science 1963; 139:1048-1049.
61. Crawford FT, Skeen LC. Operant responding in planarian—A replication study. Psychol Rep 1967; 20(3P2):1023- and.
62. Wells PH. Training flatworms in a Van Oye maze. In: Ratner WCCSC, ed. Chemistry of Learning. New York: Plenum, 1967:251-254.
63. Wells PH, Jennings LB, Davis M. Conditioning planarian worms in a Van Oye type maze. Am Zool 1966; 6(3):295.
64. Morgan T. Regeneration. New York: Macmillan, 1901.
65. Algeri S, Carolei A, Ferretti P et al. Effects of dopaminergic agents on monoamine levels and motor behaviour in planaria. Comp Biol Physiol—C: Comp Biol Physiol 1983; 74(1):27-29.
66. Hinrichs MA. A study of the physiological effects of caffeine upon planaria dorotocephala. J Exp Zool 1923; 40(2):271-300.
67. Needleman HL. Tolerance and dependence in the planarian after continuous exposure to morphine. Nature 1967; 215(102):784-785.
68. Raffa RB, Dasrath CS, Brown DR. Disruption of a drug-induced choice behavior by UV light. Behav Pharmacol 2003; 14(7):569-571.
69. Buttarelli FR, Pontieri FE, Margotta V et al. Acetylcholine/dopamine interaction in planaria. Comp Biol Physiol Toxicol Pharmacol: Cbp 2000; 125(2):225-231.
70. Kusayama T, Watanabe S. Reinforcing effects of methamphetamine in planarians. Neuroreport 2000; 11(11):2511-2513.
71. Newmark PA, Alvarado AS. Not your father's planarian: a classic model enters the era of functional genomics. Nat Rev Gen 2002; 3(3):210-219.
72. Sanchez Alvarado A, Newmark PA, Robb SM et al. The Schmidtea mediterranea database as a molecular resource for studying platyhelminthes, stem cells and regeneration. Development 2002; 129(24):5659-5665.
73. Nishimura K, Kitamura Y, Inoue T et al. Reconstruction of dopaminergic neural network and locomotion function in planarian regenerates. Dev Neurobiol 2007; 67(8):1059-1078.
74. Robb SM, Ross E, Alvarado AS. SmedGD: the Schmidtea mediterranea genome database. Nucleic Acids Res 2008; 36(Database issue):D599-606.
75. Pagan OR, Rowlands AL, Azam M et al. Reversal of cocaine-induced planarian behavior by parthenolide and related sesquiterpene lactones. Pharmacol Biochem Behav 2007 in press.
76. Nielsen DM, Crnic LS. Automated analysis of foot-shock sensitivity and concurrent freezing behavior in mice. J Neurosci Methods 2002; 115(2):199-209.
77. Boisvert MJ, Sherry DF. A system for the automated recording of feeding behavior and body weight. Physiol Behav 2000; 71(1-2):147-151.
78. Valentinuzzi VS, Kolker DE, Vitaterna MH et al. Automated measurement of mouse freezing behavior and its use for quantitative trait locus analysis of contextual fear conditioning in (BALB/cJ x C57BL/6J) F2 mice. Learn Mem 1998; 5(4-5):391-403.
79. Madrid JA, Matas P, Sanchez-Vazquez FJ et al. A contact eatometer for automated continuous recording of feeding behavior in rats. Physiol Behav 1995; 57(1):129-134.
80. Hulsey MG, Martin RJ. A system for automated recording and analysis of feeding behavior. Physiol Behav 1991; 50(2):403-408.
81. Sanberg PR, Hagenmeyer SH, Henault MA. Automated measurement of multivariate locomotor behavior in rodents. Neurobehav Toxicol Teratol 1985; 7(1):87-94.
82. Torello MW, Czekajewski J, Potter EA et al. An automated method for measurement of circling behavior in the mouse. Pharmacol Biochem Behav 1983; 19(1):13-17.
83. Fernandez de Miguel F, Cohen J, Zamora L et al. An automated system for detection and analysis of locomotor behavior in crustaceans. Bol Estud Med Biol 1989; 37(3-4):71-76.
84. McConnell JV, Cornwell PR, Clay M. An apparatus for conditioning planaria. Am J Psychol 1960; 73(4):618-622.
85. Sadauskas KK, Shuranova Zh P. (Method of recording the actograms of small aquatic animals and primary automatic processing of the information). Zh Vyssh Nerv Deiat Im I P Pavlova 1982; 32(6):1176-1179.
86. Cronin CJ, Mendel JE, Mukhtar S et al. An automated system for measuring parameters of nematode sinusoidal movement. BMC Genet 2005; 6(1):5.
87. Hicks C, Sorocco D, Levin M. Automated analysis of behavior: A computer-controlled system for drug screening and the investigation of learning. J Neurobiol 2006; 66(9):977-990.

CHAPTER 12

The Planarian Regeneration Model as a Context for the Study of Drug Effects and Mechanisms

Néstor J. Oviedo and Michael Levin*

Abstract

A complete understanding of drug effects and the mechanisms of addiction include a molecular characterization of changes in neurotransmitter and ion channel pathways and their consequences for remodeling of the nervous system and the control of adult stem and terminal somatic cell populations. The planarian offers unprecedented advantages in this area because of a powerful suite of physiological, behavioral, cell-biological and genetic tools, as well as an accessible and plentiful adult stem cell population. Most crucially, planarians possess powerful regenerative abilities, able to restore the entire body (including brain and peripheral nervous systems) from small fragments of the animal. On-going and future work on the role of ion flows, neurotransmitters and other mechanisms controlling regenerative pattern and adult remodeling will allows insight into the most crucial aspects of drug action on the brain and other organ systems.

Introduction to Planarian Regeneration

Planarians are nonparasitic invertebrates and members of the phylum platyhelminthes (flatworms). They are bilaterally symmetrical organisms, between 1-30 mm in length (under laboratory conditions), that possess a centralized nervous system and different tissues that originate from three germ layers (ectoderm, mesoderm and endoderm). Although the exact phyletic relationship between platyhelminthes and other non-ecdysozoan protostomes remains to be elucidated, key features are preserved between flatworms and vertebrate animals (e.g., bilateral symmetry, cephalization and dorsoventral and anteroposterior polarities).[1-3]

Planarians are fairly common animals found in soil, sea and freshwater habitats. While hundreds of planarian species exist, this discussion refers only to freshwater planarians, as this is the most common type used in laboratories around the world. Most freshwater planarians are easily reared and maintained under controlled conditions in the laboratory. Planarians can reproduce sexually and asexually or mixed (both sexual/asexual). Those that reproduce by sexual means are hermaphroditic but curiously need cross-fertilization in order to produce progeny (i.e., eggs from one planarian can only be fertilized by the sperm of another worm). The embryonic development of freshwater planarians was documented in a number of classic works mainly from the last century and although some modern attempts have been performed,[3-6] mechanistic process of planarian development remains mostly unknown. Asexual reproduction in planarians is by transverse fission in which the worm basically splits its body in two. The resulting fragments regenerate the

*Corresponding Author: Michael Levin—The Forsyth Institute and Department of Developmental Biology, Harvard School of Dental Medicine, 140 The Fenway, Boston, Massachusetts 02115, USA. Email: mlevin@forsyth.org

Planaria: A Model for Drug Action and Abuse, edited by Robert B. Raffa and Scott M. Rawls.

missing parts (head/tail) in about one week; thus, two new animals result from this clonal event. Importantly, both sexual and asexual reproductions are influenced by intrinsic (e.g., length of the worm, metabolic status etc.) and extrinsic factors (e.g., number of animals, temperature, dark/light cycle, food availability etc).

Planarians are well known for their extraordinary adult plasticity and regenerative capacities that have attracted the attention of researchers for over two hundred years[7] as a system that can teach developmental biologists profound lessons about the maintenance and restoration of complex structures by living systems (Fig. 1). Adult planarians can regulate their body size depending on food availability (Fig. 2). In the presence of plentiful nutrients, they grow to a determined size; however, when starved for several weeks or months, their body length is reduced and they shrink allometrically. Both growth and "degrowth" (reduction of body length) are associated with a well-regulated and controlled addition and subtraction of cells respectively.[8-10] Thus, controlling the metabolic status of animal used during experimentation is required.

The incredible regenerative powers of planarians are illustrated by their capacity to regenerate entire animals from tiny fragments (approximately 1/300th of their body or tissue fragments with at least 10,000 cells).[11,12] Both adult plasticity and regenerative capacity observed in adult planarians rely on an undifferentiated cell population that gives rise to any tissue type in their body. These undifferentiated cells are known as neoblasts[7] and are scattered throughout the body. In amputated worms, neoblasts respond to wound signals by proliferation, migration and differentiation, giving rise to the missing tissue. In intact worms the neoblasts divide, giving rise to progeny that maintain senescing differentiated tissues as well as new stem cells that maintain the undifferentiated population.

Regeneration as a Unique Context for the Study of Drug Effects and Addiction

The process of regeneration in planaria is a unique model system for the broader understanding of neuroactive drug effects on biological systems. It offers not only the potential of molecular investigation of drug effects on a complex and plastic neural network but also, due to the presence of a large and well-characterized stem cell population, the opportunity to investigate the effects of drugs on neural (and nonneural) stem cells.[13-15] The potential of the planarian model system for study of addiction and drug action arises through three main avenues of research.

First, it is now known that considerable remodeling of neural structures can take place following the use of various drugs.[16] The planarian offers an extreme example of whole-body remodeling following amputation or growth/degrowth. Indeed, recent work has identified a tight linkage between the remodeling of gene expression of putative peripheral neurons in the head and the allometric proportions of the rest of the body.[9] Another recent study has begun to characterize, at the molecular level, the reconstruction of dopaminergic neural networks and locomotor behavior in planarians during regeneration.[17] Thus, an understanding of the molecular mechanisms that allow changes in body shape and size to be transduced into precise alterations of the number and connectivity of neural paths will shed important light on the native and drug-induced changes in the structure of the brain and peripheral nervous systems. This information would be crucial to developing techniques to protect the brain from and reverse, pathological remodeling taking place as a result of addiction or drug use.

Second, it is now known that some neurotransmitters have effects on nonneural cell behavior. For example, serotonin is known to modulate the processes establishing heart and craniofacial morphogenesis.[18-28] Recent molecular data revealed that serotonin and 5HT receptors R3 and R4, as well as the transporter SERT, are utilized as a prenervous morphogen during left-right patterning and controls the large-scale asymmetry of the vertebrate bodyplan by mediating sidedness decisions in cells long prior to the development of the nervous system.[29-32] Since drug use alters numerous neurotransmitter secretion, uptake and receptor pathways, it is clear that these changes may have important implications for numerous subsystems in the adult organism as well as during embryonic development. The planarian is an ideal context in which to understand nonneural

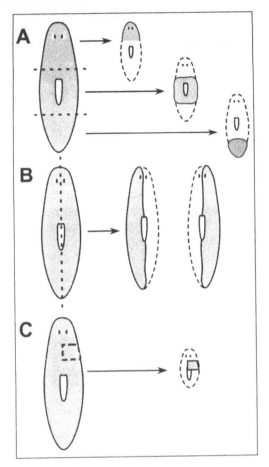

Figure 1. Planarians' regenerative capacity.Planarians can be cut in different planes (longitudinal, transversal, etc)and a new animal is recreated from pre-existing tissue. In planarians, the process of regeneration involves both active cell proliferation (epimorphosis) and remodeling of pre-existing tissue (morphallaxis). A Worms amputated transversally at two different levels (pre and post-pharyngeal levels, dashed lines) produce three fragments (head, trunk and tail) that each regenerate the missing parts in about one week. B) Longitudinal amputation across the midline produces two mirror fragments (left and right). Each fragment will also proportionally recreate the missing tissue. C) Small fragments are also able to regenerate the entire animal in short period of time. In all cases, the final animal is recreated based on both process of epimorphosis and morphollaxis represented above with white and diagonal bars from pre-existing original fragment which is illustrated in gray. Notice that despite pre-existing fragments' being located at different levels along the antero-posterior axis, they are able to create posterior, anterior/posterior, and anterior functional tissues that proportionally match their new size. This is a morphogenetic system that can recognize damage, execute the necessary repairs to restore the target morphology, and stop when it is complete. In all cases anterior is to the top.

consequences of drug-modulated neurotransmitter activity because regeneration morphology is a sensitive indicator of changes in cell number, type, or behavior following amputation and exposure to drugs. Indeed this has already begun to be investigated (reviewed in ref. 33).

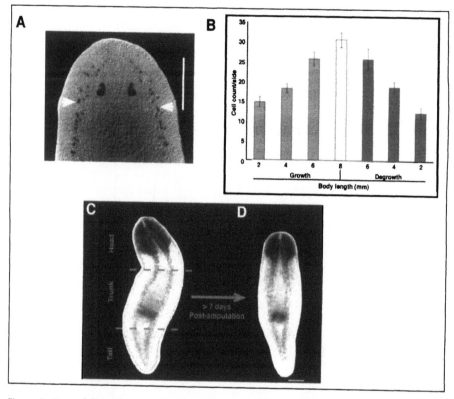

Figure 2. Remodeling of the nervous system in *S. mediterranea* during regeneration. Gene expression visualized by whole-mount in situ hybridization is used to monitor changes in the nervous system during growth/degrowth and regeneration. A) Representative expression pattern of the gene *cintillo* (a homologue of a sodium channel) in putative chemoreceptor neurons (purple precipitation limited to the anterior dorsal margin of the animal, white arrows). B) As the *cintillo* expression is restricted to a subset of cells, it is easily quantifiable under different conditions (e.g., growth and degrowth—when animals are fed or starved respectively). Notice that in both cases growth and degrowth (green and red bars respectively), the number of *cintillo* expressing cells changes proportionally with the size of the worm, suggesting that components of the nervous system are adjusted to match the total length of the animal. Results are average number of cells per side (left or right) expressing *cintillo* during growth or degrowth—as planarians increase/decrease mm in length as a consequence of feeding or starving. The yellow bar shows the combined number of cells per side expressing *cintillo* at 8 mm of length in samples at the start of degrowth conditions and samples at the end of the growth conditions (P 0.91). *cintillo* quantification was taken from Oviedo et al, 2003 with permission. C) Representative image of the expression of *smedinx-3* (a gap junction gene expressed in the planarian CNS and peripheral neurons, purple precipitation).[46] Notice that *smedinx-3* expression in the anterior end (top) is restricted to the brain (inverted "U" shape) and is also expressed in two parallel lines traveling along the antero-posterior axis (ventral nerve cords) and in a nervous plexus in the pharynx (transversal precipitation approximately in the middle of the animal). D) Representative image of *smedinx-3* expression in an animal that regenerates from the trunk fragment about one week after amputation. It reveals that the nervous system (brain and ventral nerve cords) is regenerated in about one week post-amputation. In all cases anterior is to the top and scale bars 100 μm. A color version of this figure is available at www.eurekah.com.

Finally, a major component of drug action in the brain takes place through changes in ion channel activity.[34-38] Data developed over the last 50 years indicate that endogenous bioelectric fields and voltage gradients are a key aspect of cellular controls functioning alongside the more familiar biochemical signaling pathways.[39-41] More recently, molecular advances have revealed that ion transport mediated by specific channels and pumps is an important epigenetic regulator of cell migration, proliferation and differentiation.[42,43] These biophysical controls function during embryonic development, neoplasm and regeneration—complex events that are central to the establishment and regulation of 3-dimensional shape. The data thus indicate that it is likely that drug-mediated changes in cells' electrical properties may have ramifications beyond the nervous system and could affect the behavior of cells in multiple tissues and organs; planarian regeneration is a sensitive system for detection and characterization of such events. Our lab has begun the investigation of the role of ion transport in planaria and have identified several molecular targets that endogenously control patterning in regenerating and intact planarians,[44] and are known to be targets of addictive drug exposure. For example, gap junctions, which establish isopotential cell fields, are crucial to proper morphogenesis[45] and neoblast function in planaria,[46] as well as being known targets of cocaine administration.[7,48] A functional schematic of the relationship between bioelectrical signals and regenerative morphology in planaria is given in Figure 3.

Evolutionary, Pharmacological and Behavioral Studies in Planarians

Planarians are amenable to pharmacological, physiological, molecular genetic and classical surgical techniques. Thus they are a popular model in state-of-the-art studies in regenerative biology, as well as a low-cost and popular system for high-school science projects (an ideal system to introduce students to the exciting field of drug effects and mechanisms). In the past, extensive research using different type of drugs has been performed with the intention of studying the response of the adult planarian to such treatments. For example, specific protocols have been used to study the effects of carcinogens on neoblast functions, morphogenesis/polarity during regeneration and animal behavior. Interestingly, the effect of several carcinogens in planarians led to the conclusion that this flatworm exhibit similar cellular responses to chemical insults as those observed in higher organisms.[49-54] Furthermore, recent use of psychiatric drugs that are known modulators of the serotonergic signaling (i.e., p-choloro-phenylalanine (PCPA) and reserpine) suggested that worms' responses to these compounds are very similar to those observed in higher animals.[55]

The pharmacological responses observed in planarians reveal the considerable evolutionary conservation, at the molecular and functional levels, of proteins and signaling pathways present in different organs and systems throughout phyla. The serotonergic system represents one interesting example of such molecular conservation that has intrigued researchers for long time.[1-3,55,56] In addition, many other neural receptors (e.g., fibroblast growth factor receptor, acetylcholine receptor, netrin receptor),[57-60] neuroactive molecules (e.g., gamma-aminobutyric acid, FMRFamide-like peptides),[61] axon guidance molecules (netrin, roundabout receptor *robo*)[2,58,62] and other components of the planarian CNS (e.g., homologues of the synaptotagmin and prohormone convertase-2 genes, presence of serotonergic and dopaminergic neurons)[59,63,64] are evolutionarily conserved. Thus, recent functional/behavioral assays for some of them suggest that this conservation holds at both structural and physiological levels.[2,55,57,60,61,63,65] This molecular conservation offers tremendous opportunities to design drug screens to identify specific targets in CNS physiology and regeneration[66] for important drugs[67] as well as to discover novel therapeutic drugs affecting known targets of interest. Thus, understanding how the CNS in these simple animals is organized and functions to provide appropriate responses to external and internal stimuli can be of further help not only to biology and the study of animal evolution but also to behavioral and regenerative medicine.

Modern Planarian Research—The Molecular Age

Recently, an increased interest in planarian biology has driven the development of molecular and genetic techniques to help dissect the biology of these fascinating organisms. Most common species currently used include those of the Dugesiidae family: *Girardia tigrina, Dugesia japonica*

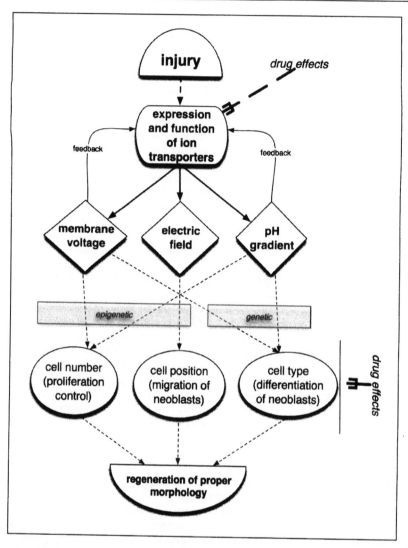

Figure 3. Logical schematic of bioelectrical signaling in planarian regeneration. Similarly to the situation in vertebrates, injury appears to trigger the up-regulation of several regeneration-specific channels and pumps. This physiological cassette alters the membrane voltage of cells that express them, produces a pH gradient and results in a long-range electric field (independent biophysical consequences of ion flux which can be perceived differentially by target cells in autonomous and non-autonomous manner). Through a variety of still largely un-characterized receptor mechanisms, these signals result in changes in cell number (control of apoptosis and proliferation), position (control of neoblast migration) and differentiation trajectory. This can occur through downstream changes of gene expression or directly through epigenetic mechanisms. The end result of this complex orchestrated cascade of biochemical, electrical and genetic factors is the rebuilding of amputated or senescing structures. Drug effects can perturb this process by direct activity on ion channel function and expression or by canonical (biochemical) controls of cell behavior. This functional network illustrates the numerous endpoints that can be used as detectors of drug activity in a regeneration assay and reveals the profound effects that drug exposure may have on pattern formation, morphostasis and physiological function within and outside the nervous system.

and *Schmidtea mediterranea.* The first is commonly found in the wild (e.g., ponds and rivers) and commercially available, while both *D. japonica* and *S. mediterranea* are from Japan and Europe/Mediterranean area respectively and have been extensively used in different laboratories (inside/outside the U.S.) to explore at molecular level themes associated with evolution, tissue regeneration and stem cell biology.[68-70]

Clonal lines has been derived from *D. japonica* and *S. mediterranea* and distributed to different laboratories around the world.[68-72] These planarians clonal lines offer less genetic variability and standardize the use of planarians research to two species. Although work from both species have contributed enormously to advance our understanding on planarian biology, recent contributions suggest that *S. mediterranea* offers more resources and the possibility of making more complete molecular analyses[70,73,74] in a stable diploid animal (2n = 8) with a genome-size that is significantly smaller than that of other planarian species.[70,71] The sequence of the genome of the diploid *S. mediterranea* is nearly completed and databases and tools are already publicly available, enabling researchers to search for genes of interest.[73-75] In addition, two different clonal strains of *S. mediterranea*—with sexual or asexual reproduction—have been established.[75,76]

Contributions from a number of labs around the world have developed and standardized cellular, molecular, genetic and genomic techniques (e.g., immunostaining, *in situ* hybridizations, RNAi, flow cytometry, DNA microarrays, etc.) that are currently being used to characterize planarian biology and evolution.[9,46,57,59,77-81] Moreover, in the last five years at least half a dozen publicly accessible planarian's databases with molecular and genetic information have been made available, facilitating the process of retrieval and comparison with other organisms' DNA sequences, gene expression profiles and RNA-interference (RNAi) phenotypes.[59,72,74-76,82-84] For the most part, current planarian work is related to genetic characterization of molecular pathways associated with tissue regeneration of different organs, stem cell (neoblast) regulation during regeneration and tissue maintenance studies.

Crucially, planarians offer a unique opportunity to study CNS regeneration in adult organisms.[45,57,58,64,66,78,85,86] Since most drug use occurs in adults, molecular investigation of the regenerative properties of the CNS and its modulation by neuroactive compounds makes planarians a very powerful model system, especially when pursued with interdisciplinary techniques including neuroscience[17,56] and automated tracking of behavior under varied experimental conditions.[55,85]

Conclusions

Planarians are becoming very popular as a model organism to dissect different areas of biology. These studies will also provide insights on how the CNS integrates and responds to different stimuli as well as to how stem cells are instructed to maintain and regenerate tissue in adult animals. The modern molecular tools now available in this animal model not only will increase our capacity to understand, prevent and correct many syndromes affecting plant, animal and humans alike but will also help to approach many neurobiological issues that are difficult to evaluate in other popular model species.

Acknowledgements

N.J.O. is a NIH fellow supported under Ruth L Kirschstein National Research Service Award (F32 GM078774). This work was also supported by NSF grant IBN#0347295, NHTSA grant DTNH22-06-G-00001 and NIH grant R21 HD055850 to M.L. This review was prepared in a Forsyth Institute facility renovated with support from Research Facilities Improvement Grant Number CO6RR11244 from the National Center for Research Resources, NIH.

References

1. Baguñà J, Riutort M. The dawn of bilaterian animals: the case of acoelomorph flatworms. Bioessays 2004; 26(10):1046-1057.
2. Cebrià F. Regenerating the central nervous system: how easy for planarians! Dev Genes Evol 2007; 217(11-12):733-748.
3. Sánchez Alvarado A. The freshwater planarian Schmidtea mediterranea: embryogenesis, stem cells and regeneration. Curr Opin Genet Dev 2003; 13(4):438-444.

4. Cardona A, Fernandez J, Solana J et al. An in situ hybridization protocol for planarian embryos: monitoring myosin heavy chain gene expression. Dev Genes Evol 2005; 215(9):482-488.
5. Cardona A, Hartenstein V, Romero R. The embryonic development of the triclad Schmidtea polychroa. Dev Genes Evol 2005; 215(3):109-131.
6. Cardona A, Hartenstein V, Romero R. Early embryogenesis of planaria: a cryptic larva feeding on maternal resources. Dev Genes Evol 2006; 216(11):667-681.
7. Reddien PW, Sánchez Alvarado A. Fundamentals of planarian regeneration. Annu Rev Cell Dev Biol 2004; 20:725-757.
8. Baguñà J, Romero R. Quantitative analysis of cell types during growth, degrowth and regeneration in the planarians Dugesia mediterranea and Dugesia tigrina. Hydrobiologia 1981; 84:181-194.
9. Oviedo NJ, Newmark PA, Sánchez Alvarado A. Allometric scaling and proportion regulation in the freshwater planarian schmidtea mediterranea. Dev Dyn 2003; 226(2):326-333.
10. Romero R, Baguñà J. Quantitative cellular analysis of growth and reproduction in freshwater planarians(Turbellaria; Tricladida) I. A cellular description of the intact organism. Invert Reprod Dev 1991; 19(2):157-165.
11. Montgomery J, Coward S. On the minimal size of a planarian capable of regeneration. Trans Am Microsc Soc 1974; 93(3):386-391.
12. Morgan TH. Experimental studies of the regeneration of Planaria maculata. Arch Entw Mech Org 1898; 7:364-397.
13. Eisch AJ, Harburg GC. Opiates, psychostimulants and adult hippocampal neurogenesis: Insights for addiction and stem cell biology. Hippocampus 2006; 16(3):271-286.
14. Yamaguchi M, Suzuki T, Seki T et al. Decreased cell proliferation in the dentate gyrus of rats after repeated administration of cocaine. Synapse 2005; 58(2):63-71.
15. Reece AS, Davidson P. Deficit of circulating stem—progenitor cells in opiate addiction: a pilot study. Subst Abuse Treat Prev Policy 2007; 2:19.
16. Ahmed SH, Lutjens R, van der Stap LD et al. Gene expression evidence for remodeling of lateral hypothalamic circuitry in cocaine addiction. Proc Natl Acad Sci USA 2005; 102(32):11533-11538.
17. Nishimura K, Kitamura Y, Inoue T et al. Reconstruction of dopaminergic neural network and locomotion function in planarian regenerates. Dev Neurobiol 2007; 67(8):1059-1078.
18. Nebigil CG, Choi DS, Dierich A et al. Serotonin 2B receptor is required for heart development. Proc Natl Acad Sci USA 2000; 97(17):9508-9513.
19. Nebigil CG, Hickel P, Messaddeq N et al. Ablation of serotonin 5-HT(2B) receptors in mice leads to abnormal cardiac structure and function. Circulation 2001; 103(24):2973-2979.
20. Lauder JM, Tamir H, Sadler TW. Serotonin and morphogenesis I. Sites of serotonin uptake and -binding protein immunoreactivity in the midgestation mouse embryo. Development Supplement 1988; 102(4):709-720.
21. Moiseiwitsch JR, Raymond JR, Tamir H et al. Regulation by serotonin of tooth-germ morphogenesis and gene expression in mouse mandibular explant cultures. Archives of Oral Biology 1998; 43(10):789-800.
22. Moiseiwitsch JR, Lauder JM. Regulation of gene expression in cultured embryonic mouse mandibular mesenchyme by serotonin antagonists. Anatomy and Embryology 1997; 195(1):71-78.
23. Whitaker-Azmitia PM, Druse M, Walker P et al. Serotonin as a developmental signal. Behavioural Brain Research 1996; 73(1-2):19-29.
24. Moiseiwitsch JR, Lauder JM. Serotonin regulates mouse cranial neural crest migration. Proceedings of the National Academy of Sciences of the United States of America 1995; 92(16):7182-7186.
25. Yavarone MS, Shuey DL, Sadler TW et al. Serotonin uptake in the ectoplacental cone and placenta of the mouse. Placenta 1993; 14(2):149-161.
26. Shuey DL, Sadler TW, Tamir H et al. Serotonin and morphogenesis. Transient expression of serotonin uptake and binding protein during craniofacial morphogenesis in the mouse. Anatomy and Embryology 1993; 187(1):75-85.
27. Buznikov GA, Lambert HW, Lauder JM. Serotonin and serotonin-like substances as regulators of early embryogenesis and morphogenesis. Cell and Tissue Research 2001; 305(2):177-186.
28. Bhasin N, Kernick E, Luo X et al. Differential regulation of chondrogenic differentiation by the serotonin2B receptor and retinoic acid in the embryonic mouse hindlimb. Dev Dyn 2004; 230(2):201-209.
29. Fukumoto T, Kema IP, Levin M. Serotonin signaling is a very early step in patterning of the left-right axis in chick and frog embryos. Curr Biol 2005; 15(9):794-803.
30. Fukumoto T, Blakely R, Levin M. Serotonin transporter function is an early step in left-right patterning in chick and frog embryos. Dev Neurosci 2005; 27(6):349-363.
31. Levin M, Buznikov GA, Lauder JM. Of minds and embryos: left-right asymmetry and the serotonergic controls of preneural morphogenesis. Dev Neurosci 2006; 28(3):171-185.

32. Esser AT, Smith KC, Weaver JC et al. Mathematical model of morphogen electrophoresis through gap junctions. Dev Dyn 2006; 235(8):2144-2159.
33. Villar D, Schaeffer DJ. Morphogenetic action of neurotransmitters on regenerating planarians—a review. Biomedical and Environmental Sciences 1993; 6(4):327-347.
34. Hu XT. Cocaine withdrawal and neuro-adaptations in ion channel function. Mol Neurobiol 2007; 35(1):95-112.
35. Karle CA, Kiehn J. An ion channel 'addicted' to ether, alcohol and cocaine: the HERG potassium channel. Cardiovasc Res 2002; 53(1):6-8.
36. Boja JW, Kopajtic TA. Ion channel inhibitors may function as potential modulators of cocaine binding. Neuropharmacology 1993; 32(3):229-234.
37. Swanson KL, Albuquerque EX. Nicotinic acetylcholine receptor ion channel blockade by cocaine: the mechanism of synaptic action. J Pharmacol Exp Ther 1987; 243(3):1202-1210.
38. Lovinger DM. 5-HT3 receptors and the neural actions of alcohols: an increasingly exciting topic. Neurochem Int 1999; 35(2):125-130.
39. Jaffe L. The role of ionic currents in establishing developmental pattern. Philosophical Transactions of the Royal Society (Series B) 1981; 295:553-566.
40. Nuccitelli R. A role for endogenous electric fields in wound healing. Curr Top Dev Biol 2003; 58:1-26.
41. Lund E. Bioelectric Fields and Growth. Austin: University of Texas Press, 1947;
42. Levin M. Large-scale biophysics: ion flows and regeneration. Trends in Cell Biology 2007; 17(6):262-271.
43. McCaig CD, Rajnicek AM, Song B et al. Controlling cell behavior electrically: current views and future potential. Physiol Rev 2005; 85(3):943-978.
44. Nogi T, Yuan YE, Sorocco D et al. Eye regeneration assay reveals an invariant functional left–right asymmetry in the early bilaterian, Dugesia japonica. Laterality 2005; 10(3):193-205.
45. Nogi T, Levin M. Characterization of innexin gene expression and functional roles of gap-junctional communication in planarian regeneration. Dev Biol 2005; 287(2):314-335.
46. Oviedo NJ, Levin M. Smedinx-11 is a planarian stem cell gap junction gene required for regeneration and homeostasis. Development 2007; 134:3121-3131.
47. McCracken CB, Hamby SM, Patel KM et al. Extended cocaine self-administration and deprivation produces region-specific and time-dependent changes in connexin36 expression in rat brain. Synapse 2005; 58(3):141-150.
48. Bennett SA, Arnold JM, Chen J et al. Long-term changes in connexin32 gap junction protein and mRNA expression following cocaine self-administration in rats. Eur J Neurosci 1999; 11(9):3329-3338.
49. Best J, Morita M. Planarians as a model system for in vitro teratogenesis studies. Teratog Carcinog Mutagen 1982; 2(3-4):277-291.
50. Foster JA. Induction of neoplasms in planarians with carcinogens. Cancer Res 1963; 23:300-303.
51. Foster JA. Malformations and lethal growths in planaria treated with carcinogens. Natl Cancer Inst Monogr 1969; 31(683):683-691.
52. Hall F, Morita M, Best J. Neoplastic Transformation in the Planarian: I. Cocarcinogenesis and Histopathology. J Exp Zool 1986; 240(2):211-227.
53. Hall F, Morita M, Best J. Neoplastic Transformation in the Planarian: II. Ultrastructure of Malignant Reticuloma. J Exp Zool 1986; 240(2):229-244.
54. Schaeffer DJ. Planarians as a model system for in vivo tumorigenesis studies. Ecotoxicol Environ Saf 1993; 25(1):1-18.
55. Hicks C, Sorocco D, Levin M. Automated analysis of behavior: a computer-controlled system for drug screening and the investigation of learning. J Neurobiol 2006; 66(9):977-990.
56. Nishimura K, Kitamura Y, Inoue T et al. Identification and distribution of tryptophan hydroxylase (TPH)-positive neurons in the planarian Dugesia japonica. Neurosci Res 2007; 59(1):101-106.
57. Cebrià F, Kobayashi C, Umesono Y et al. FGFR-related gene nou-darake restricts brain tissues to the head region of planarians. Nature 2002; 419(6907):620-624.
58. Cebrià F, Newmark PA. Planarian homologs of netrin and netrin receptor are required for proper regeneration of the central nervous system and the maintenance of nervous system architecture. Development 2005; 132(16):3691-3703.
59. Mineta K, Nakazawa M, Cebria F et al. Origin and evolutionary process of the CNS elucidated by comparative genomics analysis of planarian ESTs. Proc Natl Acad Sci USA 2003; 100(13):7666-7671.
60. Ribeiro P, El-Shehabi F, Patocka N. Classical transmitters and their receptors in flatworms. Parasitology 2005; 131(Suppl):S19-40.
61. McVeigh P, Kimber MJ, Novozhilova E et al. Neuropeptide signalling systems in flatworms. Parasitology 2005; 131(Suppl):S41-55.

62. Cebria F, Newmark PA. Morphogenesis defects are associated with abnormal nervous system regeneration following roboA RNAi in planarians. Development 2007; 134(5):833-837.
63. Agata K, Soejima Y, Kato K et al. Structure of the planarian central nervous system (CNS) revealed by neuronal cell markers. Zool Sci 1998; 15:433-440.
64. Cebrià F, Kudome T, Nakazawa M et al. The expression of neural-specific genes reveals the structural and molecular complexity of the planarian central nervous system. Mech Dev 2002; 116(1-2):199-204.
65. Saitoh O, Yuruzume E, Nakata H. Identification of planarian serotonin receptor by ligand binding and PCR studies. Neuroreport 1996; 8(1):173-178.
66. Inoue T, Hayashi T, Takechi K et al. Clathrin-mediated endocytic signals are required for the regeneration of, as well as homeostasis in, the planarian CNS. Development 2007; 134(9):1679-1689.
67. Adams DS, Levin M. Inverse drug screens: a rapid and inexpensive method for implicating molecular targets. Genesis 2006; 44(11):530-540.
68. Agata K, Watanabe K. Molecular and cellular aspects of planarian regeneration. Semin Cell Dev Biol 1999; 10(4):377-383.
69. Saló E. The power of regeneration and the stem-cell kingdom: freshwater planarians (Platyhelminthes). Bioessays 2006; 28(5):546-559.
70. Sánchez Alvarado A. Planarian regeneration: its end is its beginning. Cell 2006; 124(2):241-245.
71. Newmark PA, Sánchez Alvarado A. Not your father's planarian: a classic model enters the era of functional genomics. Nat Rev Genet 2002; 3(3):210-219.
72. Rossi L, Salvetti A, Batistoni R et al. Planarians, a tale of stem cells. Cell Mol Life Sci 2008; 65(1):16-23.
73. Cantarel BL, Korf I, Robb SM et al. MAKER: An easy-to-use annotation pipeline designed for emerging model organism genomes. Genome Res 2007; 18(1):188-196.
74. Robb SM, Ross E, Sánchez Alvarado A. SmedGD: the Schmidtea mediterranea genome database. Nucleic Acids Res 2007; 36(Database issue):D599-D606.
75. Zayas RM, Hernandez A, Habermann B et al. The planarian Schmidtea mediterranea as a model for epigenetic germ cell specification: analysis of ESTs from the hermaphroditic strain. Proc Natl Acad Sci USA 2005; 102(51):18491-18496.
76. Sánchez Alvarado A, Newmark PA, Robb SM et al. The Schmidtea mediterranea database as a molecular resource for studying platyhelminthes, stem cells and regeneration. Development 2002; 129(24):5659-5665.
77. Gonzalez-Estevez C, Felix DA, Aboobaker AA et al. Gtdap-1 promotes autophagy and is required for planarian remodeling during regeneration and starvation. Proc Natl Acad Sci USA 2007; 104(33):13373-13378.
78. Takano T, Pulvers JN, Inoue T et al. Regeneration-dependent conditional gene knockdown (Readyknock) in planarian: demonstration of requirement for Djsnap-25 expression in the brain for negative phototactic behavior. Dev Growth Differ 2007; 49(5):383-394.
79. Sánchez Alvarado A, Newmark PA. Double-stranded RNA specifically disrupts gene expression during planarian regeneration. Proc Natl Acad Sci USA 1999; 96(9):5049-5054.
80. Reddien PW, Oviedo NJ, Jennings JR et al. SMEDWI-2 is a PIWI-like protein that regulates planarian stem cells. Science 2005; 310:1327-1330.
81. Palakodeti D, Smielewska M, Graveley BR. MicroRNAs from the planarian schmidtea mediterranea: a model system for stem cell biology. Rna 2006; 12(9):1640-1649.
82. Rossi L, Salvetti A, Marincola FM et al. Deciphering the molecular machinery of stem cells: a look at the neoblast gene expression profile. Genome Biol 2007; 8(4):R62.
83. Ishizuka H, Maezawa T, Kawauchi J et al. The Dugesia ryukyuensis database as a molecular resource for studying switching of the reproductive system. Zoolog Sci 2007; 24(1):31-37.
84. Reddien PW, Bermange AL, Murfitt KJ et al. Identification of genes needed for regeneration, stem cell function and tissue homeostasis by systematic gene perturbation in planaria. Dev Cell 2005; 8(5):635-649.
85. Inoue T, Kumamoto H, Okamoto K et al. Morphological and functional recovery of the planarian photosensing system during head regeneration. Zoolog Sci 2004; 21(3):275-283.
86. Cebrià F, Nakazawa M, Mineta K et al. Dissecting planarian central nervous system regeneration by the expression of neural-specific genes. Dev Growth Differ 2002; 44(2):135-146.

APPENDIX

Bibliography of Planarian Literature

Aeppli E. [Natural polyploidia in planaria Dendrocoelum lacteum (Mueller) and Dendrocoelum infernale (Steinmann).]. Z Indukt Abstamm Vererbungsl 1952;84(3):182-212.

Agata K. The Zoological Society Prize. Molecular and cellular approaches to planarian regeneration. Zoolog Sci 2002;19(12):1391-1392.

Agata K. Regeneration and gene regulation in planarians. Curr Opin Genet Dev 2003;13(5):492-496.

Agata K. [Molecular approach to planarian stem cell system]. Tanpakushitsu Kakusan Koso 2005;50(6 Suppl):706-710.

Agata K, Tanaka T, Kobayashi C, Kato K, Saitoh Y. Intercalary regeneration in planarians. Dev Dyn 2003;226(2):308-316.

Agata K, Umesono Y. [Evolution of the genetic program controlling brain development]. Tanpakushitsu Kakusan Koso 1999;44(3):245-249.

Agata K, Watanabe K. Molecular and cellular aspects of planarian regeneration. Semin Cell Dev Biol 1999;10(4):377-383.

Algeri S, Carolei A, Ferretti P, Gallone C, Palladini G, Venturini G. Effects of dopaminergic agents on monoamine levels and motor behaviour in planaria. Comp Biochem Physiol C 1983;74(1):27-29.

Allen WB, Nollen PM. A comparative study of the regenerative processes in a trematode, Philophthalmus gralli, and a planarian, Dugesia dorotocephala. Int J Parasitol 1991;21(4):441-447.

Alonso A, Camargo JA. Toxicity of nitrite to three species of freshwater invertebrates. Environ Toxicol 2006;21(1):90-94.

Anisimov VN. Effects of exogenous melatonin--a review. Toxicol Pathol 2003;31(6):589-603.

Arees EA. Absence of light response in eyeless planaria. Physiol Behav 1986;36(3):445-449.

Arrabal PM, Estivill-Torrus G, Miranda E, Perez J, Fernandez-Llebrez P. Identification of Reissner's fiber-like glycoproteins in two species of freshwater planarians (Tricladida), by use of specific polyclonal and monoclonal antibodies. Cell Tissue Res 2000;300(3):427-434.

Arru G, Congiu AM, Burdino E, Ugazio G. [Toxicity of atrazine and its metabolite deethy-latrazine in Thamnocephalus platyurus and Dugesia gonocephala]. G Ital Med Lav Ergon 1997;19(1):17-19.

Asada A, Kusakawa T, Orii H, Agata K, Watanabe K, Tsubaki M. Planarian cytochrome b(561): conservation of a six transmembrane structure and localization along the central and peripheral nervous system. J Biochem (Tokyo) 2002;131(2):175-182.

Asada A, Orii H, Watanabe K, Tsubaki M. Planarian peptidylglycine-hydroxylating monooxy-genase, a neuropeptide processing enzyme, colocalizes with cytochrome b561 along the central nervous system. Febs J 2005;272(4):942-955.

Planaria: A Model for Drug Action and Abuse, edited by Robert B. Raffa and Scott M. Rawls.
©2008 Landes Bioscience.

Asami M, Nakatsuka T, Hayashi T, Kou K, Kagawa H, Agata K. Cultivation and characterization of planarian neuronal cells isolated by fluorescence activated cell sorting (FACS).
Zoolog Sci 2002;19(11):1257-1265.

Autuori F, Nardelli MB, Gabriel A. [Proteolytic Enzymes and Phosphatases During
Regeneration of Planaria Fragments.]. C R Hebd Seances Acad Sci 1965;260:995-998.

Azuma K. Light-induced extracellular changes of calcium and sodium concentrations in the
Planarian ocellus. Comp Biochem Physiol A Mol Integr Physiol 1998;119(1):321-325.

Azuma K, Okazaki Y, Asai K, Iwasaki N. Electrical responses and K+ activity changes to
light in the ocellus of the planarian Dugesia japonica. Comp Biochem Physiol A Physiol
1994;109(3):593-599.

Baguna J. Dramatic mitotic response in planarians after feeding, and a hypothesis for the control mechanism. J Exp Zool 1974;190(1):117-122.

Baguna J, Salo E, Romero R. Effects of activators and antagonists of the neuropeptides substance P and substance K on cell proliferation in planarians. Int J Dev Biol
1989;33(2):261-266.

Balavoine G. Identification of members of several homeobox genes in a planarian using a ligation-mediated polymerase chain reaction technique. Nucleic Acids Res
1996;24(8):1547-1553.

Balavoine G, Telford MJ. Identification of planarian homeobox sequences indicates
the antiquity of most Hox/homeotic gene subclasses. Proc Natl Acad Sci U S A
1995;92(16):7227-7231.

Baldwin RL, Wells MR. Effect of DDT on NADH-cytochrome b5 reductase activity in the
freshwater planarian, Phagocata velata. Bull Environ Contam Toxicol 1978;19(4):428-430.

Barnes CD, Katzung BG. Stimulus polarity and conditioning in planaria. Science
1963;141:728-730.

Bartnik E, Osborn M, Weber K. Intermediate filaments in muscle and epithelial cells of nematodes. J Cell Biol 1986;102(6):2033-2041.

Batey BH, Wells MR. Effects of p,p'-DDT, p,p'-DDD, and p,p'-DDE on oxygen uptake in the freshwater planarian (Phagocata gracilis). Bull Environ Contam Toxicol
1980;24(1):128-133.

Bautz A, Richoux JP, Schilt J. Demonstration by immunocytochemical staining of a soma-
tostatin-28-(1-14)-like peptide in planarians (Plathyhelminthes Turbellaria Tricladida). Gen
Comp Endocrinol 1990;78(3):469-473.

Bautz A, Schilt J. Somatostatin-like peptide and regeneration capacities in planarians. Gen
Comp Endocrinol 1986;64(2):267-272.

Bayascas JR, Castillo E, Munoz-Marmol A, Salo E. Hox genes disobey colinearity and do
not distinguish head from tail during planarian regeneration. Int J Dev Biol 1996;Suppl
1:173S-174S.

Bayascas JR, Castillo E, Munoz-Marmol AM, Salo E. Planarian Hox genes: novel patterns of
expression during regeneration. Development 1997;124(1):141-148.

Becker-Carus C. [Circadian periodicity of adaptation behavior and selection activity in pla-
naria]. Naturwissenschaften 1969;56(8):426.

Becker-Carus C. [Inhibitory mechanisms in the learning behavior of brook planaria].
Naturwissenschaften 1969;56(5):288.

Benazzi M. [Genetic and karyologic research on the planaria Dugesia lugubris (O. Schmidt)].
Boll Soc Ital Biol Sper 1952;28(4):673-674.

Benazzi M. [Results of the crossing of karyologic biotypes of the planaria Dugesia lugubris.].
Boll Soc Ital Biol Sper 1953;29(3):305-306.

Benazzi M, Lentati GB. [Gynogenesis and polyploidy in the animal kingdom]. Riv Biol
1999;92(3):452-454.

Berkowitz GC, Tschirgi RD. On the phylogeny of asymmetry and spatial discrimination. J Theor Biol 1984;106(4):495-528.

Bernardo-Filho M, Pires ET, Boasquevisque EM, Hasson-Voloch A. Labelling of the platyhelminth Dugesia tigrina with 99mtechnetium. Braz J Med Biol Res 1989;22(6):787-789.

Bessho Y, Ohama T, Osawa S. Planarian mitochondria. II. The unique genetic code as deduced from cytochrome c oxidase subunit I gene sequences. J Mol Evol 1992;34(4):331-335.

Bessho Y, Ohama T, Osawa S. Planarian mitochondria. I. Heterogeneity of cytochrome c oxidase subunit I gene sequences in the freshwater planarian, Dugesia japonica. J Mol Evol 1992;34(4):324-330.

Bessho Y, Tamura S, Hori H, Tanaka H, Ohama T, Osawa S. Planarian mitochondria sequence heterogeneity: relationships between the type of cytochrome c oxidase subunit I gene sequence, karyotype and genital organ. Mol Ecol 1997;6(2):129-136.

Best JB. Diurnal cycles and cannibalism in planaria. Science 1960;131:1884-1885.

Best JB. Studies on the Mechanisms of Memory and Morphogenesis Using Planaria as a Model System. Bol Inst Estud Med Biol Univ Nac Auton Mex 1964;22:177-190.

Best JB, Abelein M, Kreutzer E, Pigon A. Cephalic mechanism for social control of fissioning in planarians: III. Central nervous system centers of facilitation and inhibition. J Comp Physiol Psychol 1975;89(8):923-932.

Best JB, Howell W, Riegel V, Abelein M. Cephalic mechanism for social control of fissioning in planarians. I. Feedback cue and switching characteristics. J Neurobiol 1974;5(5):421-442.

Best JB, Morita M. Planarians as a model system for in vitro teratogenesis studies. Teratog Carcinog Mutagen 1982;2(3-4):277-291.

Best JB, Morita M, Abbotts B. Acute toxic responses of the freshwater planarian, Dugesia dorotocephala, to chlordane. Bull Environ Contam Toxicol 1981;26(4):502-507.

Best JB, Morita M, Ragin J, Best J, Jr. Acute toxic responses of the freshwater planarian, Dugesia dorotocephala, to methylmercury. Bull Environ Contam Toxicol 1981;27(1):49-54.

Best JB, Rosenvold R, Souders J, Wade C. Studies on the incorporation of isotopically labeled nucleotides and amino acids in planaria. J Exp Zool 1965;159(3):397-403.

Best JB, Rubinstein I. Maze learning and associated behavior in planaria. J Comp Physiol Psychol 1962;55:560-566.

Betchaku T. A copper sulfate-silver nitrate method for nerve fibers of planarians. Stain Technol 1960;35:215-218.

Betchaku T. The cellular mechanism of the formation of a regeneration blastema of fresh-water planaria, Dugesia dorotocephala. I. The behavior of cells in a tiny body fragment isolated in vitro. J Exp Zool 1970;174(3):253-279.

Bianki VL, Sheiman IM, Zubina EV. [The preference for movement direction in a T-maze in planarians]. Zh Vyssh Nerv Deiat Im I P Pavlova 1990;40(1):102-107.

Blackstone NW. Charles Manning Child (1869-1954): the past, present, and future of metabolic signaling. J Exp Zoolog B Mol Dev Evol 2006;306(1):1-7.

Bogdanova E, Matz M, Tarabykin V, Usman N, Shagin D, Zaraisky A, Lukyanov S. Inductive interactions regulating body patterning in planarian, revealed by analysis of expression of novel gene scarf. Dev Biol 1998;194(2):172-181.

Bogdanova EA, Barsova EV, Pun'kova NI, Britanova OV, Shagin DA, Gurskaia NG, Usman N, Luk'ianov SA. [A family of genes of multidomain free lectins from a planarian: structure, expression, and use as markers for regeneration monitoring]. Bioorg Khim 2004;30(6):626-637.

Bogdanova EA, Matts MV, Tarabykin VS, Usman N, Luk'ianov SA. [Differential gene expression during the reparative regeneration of differing polarities in planarians]. Ontogenez 1997;28(2):132-137.

Bogorovskaia GI. [Nervous system regeneration in the planaria]. Tsitologiia 1969;11(8):964-972.

Bonaventure N. [Galvanotropism of monster regenerates of Planaria; bifid monsters & heteromorphoses.]. C R Seances Soc Biol Fil 1957;151(3):598-602.

Bonner JC, Wells MR. Comparative acute toxicity of DDT metabolites among American and European species of planarians. Comp Biochem Physiol C 1987;87(2):437-438.

Bowen ED, Ryder TA, Dark C. The effects of starvation on the planarian worm Polycelis tenuis Iijima. Cell Tissue Res 1976;169(2):193-209.

Bowen ID, den Hollander JE, Lewis GH. Cell death and acid phosphatase activity in the regenerating planarian Polycelis tenuis Iijima. Differentiation 1982;21(3):160-167.

Bowen ID, Ryder TA. Cell autolysis and deletion in the planarian Polycelis tenuis Iijima. Cell Tissue Res 1974;154(2):265-274.

Bowen ID, Ryder TA. Use of the p-nitrophenyl phosphate method for the demonstration of acid phosphatase during starvation and cell autolysis in the planarian Polycelis tenuis Iijima. Histochem J 1976;8(3):319-329.

Bowen ID, Ryder TA, Winters C. The distribution of oxidizable mucosubstances and polysaccharides in the planarian Polycelis tenuis Iijima. Cell Tissue Res 1975;161(2):263-275.

Brown FA, Jr., Park YH. A persistent monthly variation in responses of planarians to light, and its annual modulation. Int J Chronobiol 1975;3(1):57-62.

Brown FA, Pary YH. Seasonal Variations in Sign and Strength of Gamma-Taxis in Planarians. Nature 1964;202:469-471.

Bruvo R, Michiels NK, D'Souza TG, Schulenburg H. A simple method for the calculation of microsatellite genotype distances irrespective of ploidy level. Mol Ecol 2004;13(7):2101-2106.

Bueno D, Baguna J, Romero R. A central body region defined by a position-specific molecule in the planarian Dugesia (Girardia) tigrina: spatial and temporal variations during regeneration. Dev Biol 1996;178(2):446-458.

Bueno D, Baguna J, Romero R. Cell-, tissue-, and position-specific monoclonal antibodies against the planarian Dugesia (Girardia) tigrina. Histochem Cell Biol 1997;107(2):139-149.

Bueno D, Fernandez-Rodriguez J, Cardona A, Hernandez-Hernandez V, Romero R. A novel invertebrate trophic factor related to invertebrate neurotrophins is involved in planarian body regional survival and asexual reproduction. Dev Biol 2002;252(2):188-201.

Burgaya F, Garcia-Fernandez J, Riutort M, Baguna J, Salo E. Structure and expression of Spk-1, an src-related gene found in the planarian Dugesia (G) tigrina. Oncogene 1994;9(4):1267-1272.

Burke WD, Singh D, Eickbush TH. R5 retrotransposons insert into a family of infrequently transcribed 28S rRNA genes of planaria. Mol Biol Evol 2003;20(8):1260-1270.

Bustuoabad OD, Fiocchi MG, Matteucci IM. [Change in the regeneration of neural structures induced by lithium chloride]. Medicina (B Aires) 1980;40(5):547-552.

Buttarelli FR, Pontieri FE, Margotta V, Palladini G. Acetylcholine/dopamine interaction in planaria. Comp Biochem Physiol C Toxicol Pharmacol 2000;125(2):225-231.

Buttarelli FR, Pontieri FE, Margotta V, Palladini G. Cannabinoid-induced stimulation of motor activity in planaria through an opioid receptor-mediated mechanism. Prog Neuropsychopharmacol Biol Psychiatry 2002;26(1):65-68.

Callaerts P, Munoz-Marmol AM, Glardon S, Castillo E, Sun H, Li WH, Gehring WJ, Salo E. Isolation and expression of a Pax-6 gene in the regenerating and intact Planarian Dugesia(G) tigrina. Proc Natl Acad Sci U S A 1999;96(2):558-563.

Callahan JL, Morris CD. Production and maintenance of large numbers of Dugesia tigrina (Turbellaria: Tricladida) for the control of mosquitoes in the field. J Am Mosq Control Assoc 1989;5(1):10-14.

Camargo JA, Ward JV. Differential sensitivity of Dugesia dorotocephala and Cheumatopsyche pettiti to water acidification: ecological implication for predator-prey interactions. Arch Environ Contam Toxicol 1992;23(1):59-63.

Carbayo F, Leal-Zanchet AM. A new species of terrestrial planarian (Platyhelminthes: Tricladida: Terricola) from south Brazil. Braz J Biol 2001;61(3):437-447.

Cardona A, Fernandez J, Solana J, Romero R. An in situ hybridization protocol for planarian embryos: monitoring myosin heavy chain gene expression. Dev Genes Evol 2005;215(9):482-488.

Carolei A, Margotta V, Palladini G. Proposal of a new model with dopaminergic-cholinergic interactions for neuropharmacological investigations. Neuropsychobiology 1975;1(6):355-364.

Caronti B, Margotta V, Merante A, Pontieri FE, Palladini G. Treatment with 6-hydroxydopamine in planaria (Dugesia gonocephala s.l.): morphological and behavioral study. Comp Biochem Physiol C Pharmacol Toxicol Endocrinol 1999;123(3):201-207.

Carranza S, Baguna J, Riutort M. Origin and evolution of paralogous rRNA gene clusters within the flatworm family Dugesiidae (Platyhelminthes, Tricladida). J Mol Evol 1999;49(2):250-259.

Carranza S, Littlewood DT, Clough KA, Ruiz-Trillo I, Baguna J, Riutort M. A robust molecular phylogeny of the Tricladida (Platyhelminthes: Seriata) with a discussion on morphological synapomorphies. Proc Biol Sci 1998;265(1396):631-640.

Cassina R, Poletti H. [Various observations on the ultrastructure of spermatozoid flagellum of planaria (gen. Dugesia)]. An Fac Med Univ Repub Montev Urug 1966;51(1):81-84.

Cebria F, Bueno D, Reigada S, Romero R. Intercalary muscle cell renewal in planarian pharynx. Dev Genes Evol 1999;209(4):249-253.

Cebria F, Kobayashi C, Umesono Y, Nakazawa M, Mineta K, Ikeo K, Gojobori T, Itoh M, Taira M, Sanchez Alvarado A, Agata K. FGFR-related gene nou-darake restricts brain tissues to the head region of planarians. Nature 2002;419(6907):620-624.

Cebria F, Kudome T, Nakazawa M, Mineta K, Ikeo K, Gojobori T, Agata K. The expression of neural-specific genes reveals the structural and molecular complexity of the planarian central nervous system. Mech Dev 2002;116(1-2):199-204.

Cebria F, Nakazawa M, Mineta K, Ikeo K, Gojobori T, Agata K. Dissecting planarian central nervous system regeneration by the expression of neural-specific genes. Dev Growth Differ 2002;44(2):135-146.

Cebria F, Newmark PA. Planarian homologs of netrin and netrin receptor are required for proper regeneration of the central nervous system and the maintenance of nervous system architecture. Development 2005;132(16):3691-3703.

Cebria F, Vispo M, Bueno D, Carranza S, Newmark P, Romero R. Myosin heavy chain gene in Dugesia (G.) tigrina: a tool for studying muscle regeneration in planarians. Int J Dev Biol 1996;Suppl 1:177S-178S.

Chandebois R. [Behavior of the epidermis in teratomorphic regenerates in marine Planaria Procerodes lobata O. Schmidt.]. C R Hebd Seances Acad Sci 1954;239(15):911-913.

Chandebois R. [On the source of regenerative histogenesis in Planaria.]. C R Hebd Seances Acad Sci 1960;251:146-148.

Chandebois R. Cell sociology: a way of reconsidering the current concepts of morphogenesis. Acta Biotheor 1976;25(2-3):71-102.

Chandebois R. Histogenesis and morphogenesis in planarian regeneration. Monogr Dev Biol 1976;11:1-182.

Chandebois R. Differentiated epidermal outgrowths in the planarian Dugesia gonocephala: a model for studying cell renewal and patterning in single-layered epithelial tissue. Exp Cell Biol 1985;53(1):46-58.

Chandebois R. The handling of half-thickness pieces in planarians: preliminary results. Monogr Dev Biol 1988;21:235-241.

Chapouthier G, Pallaud B, Ungerer A. [Relation between 2 reactions of Planaria facing right-left discrimination]. C R Acad Sci Hebd Seances Acad Sci D 1968;266(9):905-907.

Chaurasia ON. Studies of a new land planarian, Orthodemus indica sp. nov. Folia Morphol (Praha) 1985;33(2):125-128.

Chaurasia ON. Studies of a new land planarian, Bipalium chhatarpurensis sp. nov. Folia Morphol (Praha) 1988;36(4):403-407.

Cherkashin AN, Sheiman IM, Bogorovskaia GI. [Conditioned reflexes in planaria and experiments with regeneration]. Zh Vyssh Nerv Deiat Im I P Pavlova 1966;16(6):1110-1112.

Cherkashin AN, Sheiman IM, Sergeeva EP. [Effect of combinations of light and electroshock on planaria]. Zh Vyssh Nerv Deiat Im I P Pavlova 1966;16(2):266-273.

Chernyshev AV, Isaeva VV, Presnov EV. [Comparative analysis of topological organization in Metazoa]. Zh Obshch Biol 2001;62(1):49-56.

Chien PK, Koopowitz H. Ultrastructure of nerve plexus in flatworms. III. The infra-epithelial nervous system. Cell Tissue Res 1977;176(3):335-347.

Christensen NO. Schistosoma mansoni: interference with cercarial host-finding by various aquatic organisms. J Helminthol 1979;53(1):7-14.

Christensen NO, Nansen P, Frandsen F. Interference with Fasciola hepatica snail finding by various aquatic organisms. Parasitology 1977;74(3):285-290.

Cobb JL, Pentreath VW. Comparison of the morphology of synapses in invertebrate and vertebrate nervous systems: analysis of the significance of the anatomical differences and interpretation of the morphological specializations. Prog Neurobiol 1978;10(4):231-252.

Cobbett P, Day TA. Functional voltage-gated Ca2+ channels in muscle fibers of the platyhelminth Dugesia tigrina. Comp Biochem Physiol A Mol Integr Physiol 2003;134(3):593-605.

Collet J, Baguna J. Optimizing a method of protein extraction for two-dimensional electrophoretic separation of proteins from planarians (Platyhelminthes, Turbellaria). Electrophoresis 1993;14(10):1054-1059.

Collins TF. Teratological research using in vitro systems. V. Nonmammalian model systems. Environ Health Perspect 1987;72:237-249.

Congiu AM, Casu S, Ugazio G. Toxicity of selenium (Na2SeO3) and mercury (HgCl2) on the planarian Dugesia gonocephala. Res Commun Chem Pathol Pharmacol 1989;66(1):87-95.

Corbridge A. Planaria, how they are effected by direct radiation and by irradiated food. J Colo Dent Assoc 1967;45(2):22-25.

Cornford EM. Schistosomatium douthitti: effects of thyroxine. Exp Parasitol 1974;36(2):210-221.

Corning WC. Evidence of Right-Left Discrimination in Planarians. J Psychol 1964;58:131-139.

Corning WC, John ER. Effect of ribonuclease on retention of conditioned response in regenerated planarians. Science 1961;134:1363-1365.

Coward SJ. The relation of surface and volume to so-called physiological gradients in planaria. Dev Biol 1968;18(6):590-601.

Coward SJ. Chromatoid bodies in somatic cells of the planarian: observations on their behavior during mitosis. Anat Rec 1974;180(3):533-545.

Coward SJ. On the occurrence and significance of annulate lamellae in gastrodermal cells of regenerating planarians. Cell Biol Int Rep 1979;3(2):101-106.

Coward SJ, Flickinger RA. Axial Patterns of Protein and Nucleic Acid Syntheses in Intact and Regenerating Planaria. Growth 1965;29:151-163.

Croft JA, Jones GH. Meiosis in Mesostoma ehrenbergii ehrenbergii. IV. Recombination nodules in spermatocytes and a test of the correspondence of late recombination nodules and chiasmata. Genetics 1989;121(2):255-262.

Csaba G. Ontogeny and phylogeny of hormone receptors. Monogr Dev Biol 1981;15:1-172.

Csaba G. Presence in and effects of pineal indoleamines at very low level of phylogeny. Experientia 1993;49(8):627-634.

Csaba G, Bierbauer J, Feher Z. Influenace of melatonin and its precursors on the pigment cells of planaria (Dugesia lugubris). Comp Biochem Physiol C 1980;67C(2):207-209.

Csaba G, Dobozy O, Darvas Z, Laszlo V, Beress L. Phylogenetic changes in sensitivity to Anemonia sulcata toxin (ATX II), and impact of first interaction with the toxin (imprinting) on later response to it. Comp Biochem Physiol C 1984;77(1):153-155.

Csaba G, Gruszczynska M. Effect of repeated treatments with sympathomimetic drugs on planarian glucose metabolism. Acta Physiol Hung 1983;61(3):141-145.

Csaba G, Kadar M. Effects of epinephrine, glucagon and insulin on glucose metabolism of planaria. Endokrinologie 1978;71(1):113-115.

Csaba G, Kadar M. The effect of sympathicomimetic agents on carbohydrate metabolism of planaria. Acta Physiol Acad Sci Hung 1979;53(3):323-326.

Csaba G, Kadar M. Analysis of the progress of hormone-receptor amplification during differentiation in regenerating planarians. Exp Cell Biol 1979;47(2):155-160.

Csaba G, Kadar M. Durable sensitization of hormone receptors during differentiation in regenerating planarians by treatment with homologous or analogous hormone molecules. Exp Cell Biol 1980;48(3):240-244.

D'Souza TG, Storhas M, Schulenburg H, Beukeboom LW, Michiels NK. Occasional sex in an 'asexual' polyploid hermaphrodite. Proc Biol Sci 2004;271(1543):1001-1007.

Daly JJ, Farris HE, Jr., Matthews HM. Pseudoparasitism of dogs and cats by the land planarian, Bipalium kewense. Vet Med Small Anim Clin 1976;71(11):1540-1542.

Daly JJ, Matthews HM, Farris HE, Jr. Evidence against gastrointestinal pseudoparasitism by the land planarian, Bipalium kewense Moseley 1878. Health Lab Sci 1977;14(1):39-43.

Darby WM, Boobar LR, Sardelis MR. A method for dispensing planaria (Dugesia dorotocephala) for mosquito control. J Am Mosq Control Assoc 1988;4(4):545-546.

Dasheiff BD, Dasheiff RM. Photonegative response in brown planaria (Dugesia tigrina) following regeneration. Ecotoxicol Environ Saf 2002;53(2):196-199.

Davison J. Population growth in planaria Dugesia tigrina (gerard). Regulation by the absolute number in the population. J Gen Physiol 1973;61(6):767-785.

Day TA, Haithcock J, Kimber M, Maule AG. Functional ryanodine receptor channels in flatworm muscle fibres. Parasitology 2000;120 (Pt 4):417-422.

de Campos-Velho NM, Lopes KA, Hauser J. Morphometry of the eyes in regenerant of genus Dugesia (Platyhelminthes, Turbellaria, Dugesiidae). Braz J Biol 2004;64(1):1-9.

de Lucca CPD, Smith DH. Use of newborn Girardia tigrina (Girard, 1850) in acute toxicity tests. Ecotoxicol Environ Saf 2001;50(1):1-3.

de P, Grain J. [On 2 Gregarina of the Monocystella genus, endoparasites of ochridian Planaria: Fonticola ochridana Stankovic and Neodendrocoelum sanctinaumi Stankovic.]. Ann Parasitol Hum Comp 1960;35:197-208.

de Souza SC, Munin E, Alves LP, Salgado MA, Pacheco MT. Low power laser radiation at 685 nm stimulates stem-cell proliferation rate in Dugesia tigrina during regeneration. J Photochem Photobiol B 2005;80(3):203-207.

Dessi Fulgheri F, Messeri P. [Use of 2 different negative reinforcements in light-darkness discrimination of planarians]. Boll Soc Ital Biol Sper 1973;49(20):1141-1145.

DiCiaula LL, Foley GL, Schaeffer DJ. Fixation and staining of planaria for histological study. Biotech Histochem 1995;70(3):119-123.

Domenici L, Gremigni V. Electron microscopical and cytochemical study of vitelline cells in the fresh-water triclad Dugesia lugubris s.l. II. Origin and distribution of reserve materials. Cell Tissue Res 1974;152(2):219-228.

Dougan PM, Mair GR, Halton DW, Curry WJ, Day TA, Maule AG. Gene organization and expression of a neuropeptide Y homolog from the land planarian Arthurdendyus triangulatus. J Comp Neurol 2002;454(1):58-64.

Draper AC, 3rd, Brewer WS. Measurement of the aquatic toxicity of volatile nitrosamines. J Toxicol Environ Health 1979;5(6):985-993.

Dubois F. [Continuity of the migration of neoblasts in the regeneration of the planaria Euplanaria lugubris.]. C R Seances Soc Biol Fil 1950;144(21-22):1545-1548.

Duma A. Activity in cAMP phosphodiesterase in the early regeneration stage of planarian Dugesia lugubris (O. Schmidt). Ultracytochemical studies. Acta Med Pol 1980;21(4):317-318.

Duma A, Moraczewski J. Ultracytochemistry of cyclic 3',5'-nucleotide phosphodiesterase activity in the planarian Dugesia lugubris (O. Schmidt). Histochemistry 1980;66(2):211-220.

Dvoriadkin VA, Besprozvannykh VV. [Systematic position and life cycle of Asymphylotrema macracetabulum comb. Nov. (Trematoda, Monorchidae)]. Parazitologiia 1985;19(5):394-398.

Edelman JR, Lin YJ. Differential staining of Dugesia tigrina sister chromatids. Cytobios 1994;78(313):123-128.

Edelman JR, Lin YJ. Heterochromatin (chromosome dots and chromocentres): key to planarian regeneration? Cytobios 1996;86(347):247-253.

Edelman JR, Lin YJ. 'Glowing' chromosomes in cells undergoing rapid division. Cytobios 2000;102(401):149-156.

Efimov NA, Sakharova N. [Simple method of straightening planaria during fixation and their histologic preparation]. Ontogenez 1976;7(5):537-539.

Egger B, Ladurner P, Nimeth K, Gschwentner R, Rieger R. The regeneration capacity of the flatworm Macrostomum lignano-on repeated regeneration, rejuvenation, and the minimal size needed for regeneration. Dev Genes Evol 2006.

Eriksson KS. Nitric oxide synthase in the pharynx of the planarian Dugesia tigrina. Cell Tissue Res 1996;286(3):407-410.

Eriksson KS, Panula P. gamma-Aminobutyric acid in the nervous system of a planarian. J Comp Neurol 1994;345(4):528-536.

Erofeeva ES, Golubev AI, Maliutina LV. [The adhesive system of the planarian Dendrocoelum lacteum: its structure and innervation]. Morfologiia 1998;114(4):93-99.

Etges FJ, Carter OS, Webbe G. Behavioral and developmental physiology of schistosome larvae as related to their molluscan hosts. Ann N Y Acad Sci 1975;266:480-496.

Fagotti A, Gabbiani G, Pascolini R, Neuville P. Multiple isoform recovery (MIR)-PCR: a simple method for the isolation of related mRNA isoforms. Nucleic Acids Res 1998;26(8):2031-2033.

Farnesi RM. [Some ultrastructural observations on the pharynx of specimens of Dugesia lugubris s. l. treated with colcemid]. Boll Soc Ital Biol Sper 1975;51(18):1177-1183.

Farnesi RM, Marinelli M, Vagnetti D. [Electron microscopic observations on the relations between neurosecretion and the cement gland in Dugesia lugubris s.l]. Boll Soc Ital Biol Sper 1975;51(6):329-334.

Farnesi RM, Tei S. Dugesia lugubris s.l. auricles: research into the ultrastructure and on the functional efficiency. Riv Biol 1980;73(1):65-77.

Fedecka-Bruner B. [Studies on the regeneration of the genital organs of the planaria Dugesia lugubris. I. Regeneration of the testes after destruction]. Bull Biol Fr Belg 1967;101(4):255-319.

Fernandes MC, Alvares EP, Gama P, Silveira M. Serotonin in the nervous system of the head region of the land planarian Bipalium kewense. Tissue Cell 2003;35(6):479-486.

Fire A. RNA-triggered gene silencing. Trends Genet 1999;15(9):358-363.

Flickinger RA. A gradient of protein synthesis in planaria and reversal of axial polarity of regenerates. Growth 1959;23:251-271.

Foster JA. Induction of neoplasms in planarians with carcinogens. Cancer Res 1963;23:300-303.

Foster JA. Malformations and lethal growths in planaria treated with carcinogens. Natl Cancer Inst Monogr 1969;31:683-691.

Fournier G, Lenicque PM, Paris MR. [Toxic effects of essential oil of Cannabis sativa L. and main constituents on planarian (Dugesia tigrina) (author's transl)]. Toxicol Eur Res 1978;1(6):385-389.

Franquinet R. [Comparative study of the cells of the freshwater planarian Polycelis tenuis (Iijima) using dissociated fragments cultivated in vitro: ultrastructural aspects and incorporation of (3H) leucine and (3H) uridine]. J Embryol Exp Morphol 1976;36(1):41-54.

Franquinet R. [The role of serotonin and catecholamines in the regeneration of the Planaria Polycelis tenvis]. J Embryol Exp Morphol 1979;51:85-95.

Franquinet R, Le Moigne A, Hanoune J. The adenylate cyclase system of planaria Polycelis tenuis: activation by serotonin and guanine nucleotides. Biochim Biophys Acta 1978;539(1):88-97.

Franquinet R, Lender T, Moigne AL. [Incorporation of 3H-5 uridine and 3H-5 leucine into cultured planarian cells, Polycelis tenuis (Iijima) in vitro]. C R Acad Sci Hebd Seances Acad Sci D 1975;280(19):2253-2256.

Franquinet R, Martelly I. Effects of serotonin and catecholamines on RNA synthesis in planarians; in vitro and in vivo studies. Cell Differ 1981;10(4):201-209.

Franquinet R, Stengel D, Hanqune J. The adenylate cyclase system in a freshwater planarian (Polycelis tenius Iijima). Comp Biochem Physiol B 1976;53(3):329-333.

Fried B, Rosa-Brunet LC. Exposure of Dugesia tigrina (Turbellaria) to cercariae of Echinostoma trivolvis and Echinostoma caproni (Trematoda). J Parasitol 1991;77(1):113-116.

Friedel T, Webb RA. Stimulation of mitosis in Dugesia tigrina by a neurosecretory fraction. Can J Zool 1979;57(9):1818-1819.

Fusaoka E, Inoue T, Mineta K, Agata K, Takeuchi K. Structure and function of primitive immunoglobulin superfamily neural cell adhesion molecules: a lesson from studies on planarian. Genes Cells 2006;11(5):541-555.

Gabriel A. [Action of actinomycin D on regeneration and the metabolism of ribonucleic acid in the planaria Dugesia gonocephala (Turbellaria, Tricladida)]. C R Acad Sci Hebd Seances Acad Sci D 1968;266(4):406-409.

Garat B, Esperon P, Martinez C, Robello C, Ehrlich R. Presence of a conserved domain of GATA transcription factors in Echinococcus granulosus. J Helminthol 1997;71(4):355-358.

Garcia-Fernandez J, Baguna J, Salo E. Planarian homeobox genes: cloning, sequence analysis, and expression. Proc Natl Acad Sci U S A 1991;88(16):7338-7342.

Garcia-Fernandez J, Baguna J, Salo E. Genomic organization and expression of the planarian homeobox genes Dth-1 and Dth-2. Development 1993;118(1):241-253.

Garcia-Fernandez J, Bayascas-Ramirez JR, Marfany G, Munoz-Marmol AM, Casali A, Baguna J, Salo E. High copy number of highly similar mariner-like transposons in planarian (Platyhelminthe): evidence for a trans-phyla horizontal transfer. Mol Biol Evol 1995;12(3):421-431.

Garcia-Fernandez J, Marfany G, Baguna J, Salo E. Infiltration of mariner elements. Nature 1993;364(6433):109-110.

Gehring WJ. New perspectives on eye development and the evolution of eyes and photoreceptors. J Hered 2005;96(3):171-184.

Gindilis V, Banikazemi M, Vyasankin A, Verlinsky O, Matveyev I, Verlinsky Y. Review: borders, patterns, and distinctive families of homeodomains. J Assist Reprod Genet 1994;11(5):244-269.

Gonzalez-Estevez C, Momose T, Gehring WJ, Salo E. Transgenic planarian lines obtained by electroporation using transposon-derived vectors and an eye-specific GFP marker. Proc Natl Acad Sci U S A 2003;100(24):14046-14051.

Goss LB, Sabourin TD. Utilization of alternative species for toxicity testing: an overview. J Appl Toxicol 1985;5(4):193-219.

Gourbault N. [New data concerning chromosomes of the American planaria Procotyla fluviatilis Leidy]. C R Acad Sci Hebd Seances Acad Sci D 1974;279(14):1171-1173.

Grasso M, Montanaro L, Quaglia A. Studies on the role of neurosecretion in the induction of sexuality in a planarian agamic strain. J Ultrastruct Res 1975;52(3):404-408.

Grebe E, Schaeffer DJ. Planarians in toxicology, standardization of a rapid neurobehavioral toxicity test using phenol in a crossover study. Bull Environ Contam Toxicol 1991;46(6):866-870.

Grebe E, Schaeffer DJ. Neurobehavioral toxicity of cadmium sulfate to the planarian Dugesia dorotocephala. Bull Environ Contam Toxicol 1991;46(5):727-730.

Gremigni V, Miceli C, Picano E. On the role of germ cells in planarian regeneration. II. Cytophotometric analysis of the nuclear Feulgen-DNA content in cells of regenerated somatic tissues. J Embryol Exp Morphol 1980;55:65-76.

Gremigni V, Miceli C, Puccinelli I. On the role of germ cells in planarian regeneration. I. A karyological investigation. J Embryol Exp Morphol 1980;55:53-63.

Gremigni V, Nigro M, Puccinelli I. Evidence of male germ cell redifferentiation into female germ cells in planarian regeneration. J Embryol Exp Morphol 1982;70:29-36.

Grudnitskii VA, Dontsova GV. Change in ATP and ADP content of fasting and feeding planarians (Polycelis nigra). Biol Bull Acad Sci USSR 1979;6(5):650-652.

Grudnitskii VA, Dontsova GV. Content of mitochondrial protein in the planarian Polycelis nigra during starvation and feeding. Biol Bull Acad Sci USSR 1979;6(4):501-504.

Guecheva T, Henriques JA, Erdtmann B. Genotoxic effects of copper sulphate in freshwater planarian in vivo, studied with the single-cell gel test (comet assay). Mutat Res 2001;497(1-2):19-27.

Guecheva TN, Erdtmann B, Benfato MS, Henriques JA. Stress protein response and catalase activity in freshwater planarian Dugesia (Girardia) schubarti exposed to copper. Ecotoxicol Environ Saf 2003;56(3):351-357.

Guimaraes JP, Klaczko LB, Hirano K, Vaz EM, Miguel MR. Reappearance of embryonal antigens in planarian regenerates. Rev Bras Pesqui Med Biol 1975;8(3-4):255-259.

Halas ES, James RL, Stone LA. Types of responses elicited in planaria by light. J Comp Physiol Psychol 1961;54:302-305.

Hall F, Morita M, Best JB. Neoplastic transformation in the planarian: II. Ultrastructure of malignant reticuloma. J Exp Zool 1986;240(2):229-244.

Hall F, Morita M, Best JB. Neoplastic transformation in the planarian: I. Cocarcinogenesis and histopathology. J Exp Zool 1986;240(2):211-227.

Hamana K, Hamana H, Shinozawa T. Alterations in polyamine levels of nematode, earthworm, leech and planarian during regeneration, temperature and osmotic stresses. Comp Biochem Physiol B Biochem Mol Biol 1995;111(1):91-97.

Hammond SM, Bernstein E, Beach D, Hannon GJ. An RNA-directed nuclease mediates post-transcriptional gene silencing in Drosophila cells. Nature 2000;404(6775):293-296.

Hansen LG, Tehseen WM, Foley GL, Schaeffer DJ. Modification by polychlorinated biphenyls (PCBs) of cadmium induced lesions in the planarian model, Dugesia dorotocephala. Biomed Environ Sci 1993;6(4):367-384.

Hase S, Kobayashi K, Koyanagi R, Hoshi M, Matsumoto M. Transcriptional pattern of a novel gene, expressed specifically after the point-of-no-return during sexualization, in planaria. Dev Genes Evol 2003;212(12):585-592.

Haslauer J. [Effect of Visible and Ultraviolet Light on the Regeneration Growth of Planaria.]. Strahlentherapie 1964;125:604-630.

Hauser J, Friedrich SM. Morphogenesis of regenerating fragments of Dugesia schubarti (Turbellaria tricladia). Exp Cell Biol 1982;50(2):61-71.

Hauser J, Silveira S, de Paula C. [Morphometric analysis by means of binocular stereomicroscopy (author's transl)]. Mikroskopie 1979;35(7-8):207-212.

Hauser PJ. [Histologic reorganization processes in the intestine of Planaria after feeding.]. Mikroskopie 1956;11(1-2):20-31.

Hay ED, Coward SJ. Fine structure studies on the planarian, Dugesia. I. Nature of the "neoblast" and other cell types in noninjured worms. J Ultrastruct Res 1975;50(1):1-21.

Hayashi S, Itoh M, Taira S, Agata K, Taira M. Expression patterns of Xenopus FGF receptor-like 1/nou-darake in early Xenopus development resemble those of planarian nou-darake and Xenopus FGF8. Dev Dyn 2004;230(4):700-707.

Hayashi T, Asami M, Higuchi S, Shibata N, Agata K. Isolation of planarian X-ray-sensitive stem cells by fluorescence-activated cell sorting. Dev Growth Differ 2006;48(6):371-380.

Hempstead PG, Regular SC, Ball IR. A method for the preparation of high-molecular-weight DNA from marine and freshwater triclads (Platyhelminthes, Turbellaria). DNA Cell Biol 1990;9(1):57-61.

Henderson RF, Eakin RE. Alteration of regeneration in planaria treated with lipoic acid. J Exp Zool 1959;141:175-190.

Henderson TR, Eakin RE. Irreversible alteration of differentiated tissues in planaria by purine analogues. J Exp Zool 1961;146:253-263.

Hori I. Possible role of rhabdite-forming cells in cellular succession of the planarian epidermis. J Electron Microsc (Tokyo) 1978;27(2):89-102.

Hori I. Regeneration of the epidermis and basement membrane of the planarian Dugesia japonica after total-body X irradiation. Radiat Res 1979;77(3):521-533.

Hori I. Structure and regeneration of the planarian basal lamina: an ultrastructural study. Tissue Cell 1979;11(4):611-621.

Hori I. Localization of newly synthesized precursors of basal lamina in the regenerating planarian as revealed by autoradiography. Tissue Cell 1980;12(3):513-521.

Hori I. Observations on planarian epithelization after wounding. J Submicrosc Cytol Pathol 1989;21(2):307-315.

Hori I. Role of fixed parenchyma cells in blastema formation of the planarian Dugesia japonica. Int J Dev Biol 1991;35(2):101-108.

Hori I. Cytological approach to morphogenesis in the planarian blastema. II. The effect of neuropeptides. J Submicrosc Cytol Pathol 1997;29(1):91-97.

Hori I, Kishida Y. Quantitative changes in nuclear pores and chromatoid bodies induced by neuropeptides during cell differentiation in the planarian Dugesia japonica. J Submicrosc Cytol Pathol 2003;35(4):439-444.

Horvat T, Kalafatic M, Kopjar N, Kovacevic G. Toxicity testing of herbicide norflurazon on an aquatic bioindicator species--the planarian Polycelis felina (Daly.). Aquat Toxicol 2005;73(4):342-352.

Hutchinson TH, Hutchings MJ, Moore KW. A review of the effects of bromate on aquatic organisms and toxicity of bromate to oyster (Crassostrea gigas) embryos. Ecotoxicol Environ Saf 1997;38(3):238-243.

Hutticher A, Kerschbaum HH, Kainz V, Bito M, Hermann A. Parvalbumin-immunoreactive proteins in the nervous system of planarians. Cell Mol Neurobiol 1995;15(4):401-410.

Hwang JS, Kobayashi C, Agata K, Ikeo K, Gojobori T. Detection of apoptosis during planarian regeneration by the expression of apoptosis-related genes and TUNEL assay. Gene 2004;333:15-25.

Hyden H, Egyhazi E, John ER, Bartlett F. RNA base ratio changes in Planaria during conditioning. J Neurochem 1969;16(5):813-821.

Inoue T, Kumamoto H, Okamoto K, Umesono Y, Sakai M, Sanchez Alvarado A, Agata K. Morphological and functional recovery of the planarian photosensing system during head regeneration. Zoolog Sci 2004;21(3):275-283.

Ishii S. The ultrastructure of the protonephridial tubules of the freshwater planarian Bdellocephala brunnea. Cell Tissue Res 1980;206(3):451-458.

Ishii S. The ultrastructure of the protonephridial flame cell of the freshwater planarian Bdellocephala brunnea. Cell Tissue Res 1980;206(3):441-449.

Ito H, Saito Y, Watanabe K, Orii H. Epimorphic regeneration of the distal part of the planarian pharynx. Dev Genes Evol 2001;211(1):2-9.

Itoh MT, Igarashi J. Circadian rhythm of serotonin levels in planarians. Neuroreport 2000;11(3):473-476.

Itoh MT, Shinozawa T, Sumi Y. Circadian rhythms of melatonin-synthesizing enzyme activities and melatonin levels in planarians. Brain Res 1999;830(1):165-173.

J., Azuma K, Shinozawa T. Rhodopsin-like proteins in planarian eye and auricle: detection and functional analysis. J Exp Biol 1998;201 (Pt 9):1263-1271.

Jarvik T, Lark KG. Characterization of Soymar1, a mariner element in soybean. Genetics 1998;149(3):1569-1574.

Jenrow KA, Smith CH, Liboff AR. Weak extremely-low-frequency magnetic fields and regeneration in the planarian Dugesia tigrina. Bioelectromagnetics 1995;16(2):106-112.

Jenrow KA, Smith CH, Liboff AR. Weak extremely-low-frequency magnetic field-induced regeneration anomalies in the planarian Dugesia tigrina. Bioelectromagnetics 1996;17(6):467-474.

Joffe BI, Solovei IV, Macgregor HC. Ordered arrangement and rearrangement of chromosomes during spermatogenesis in two species of planarians (Plathelminthes). Chromosoma 1998;107(3):173-183.

Joffee BI, Solovei IV, Macgregor HC. Ends of Chromosomes in Polycelis tenuis (Platyhelminthes) have telomere repeat TTAGGG. Chromosome Res 1996;4(4):323-324.

Johnson EM. A review of advances in prescreening for teratogenic hazards. Prog Drug Res 1985;29:121-154.

Johnson LR, Davenport R, Balbach H, Schaeffer DJ. Phototoxicology. 3. Comparative toxicity of trinitrotoluene and aminodinitrotoluenes to Daphnia magna, Dugesia dorotocephala, and sheep erythrocytes. Ecotoxicol Environ Saf 1994;27(1):34-49.

Johnston RN, Shaw C, Halton DW, Verhaert P, Baguna J. GYIRFamide: a novel FMRFamide-related peptide (FaRP) from the triclad turbellarian, Dugesia tigrina. Biochem Biophys Res Commun 1995;209(2):689-697.

Kalafatic M, Kopjar N, Besendorfer V. The impairments of neoblast division in regenerating planarian Polycelis felina (Daly.) caused by in vitro treatment with cadmium sulfate. Toxicol In Vitro 2004;18(1):99-107.

Kalafatic M, Kopjar N, Zrna G, Zupan I, Kovacevic G, Franjevic D. Karyological analysis of two allopatric populations of planarian Polycelis felina (Daly.) in Croatia. Folia Biol (Krakow) 2004;52(1-2):131-133.

Kamliuk LV. [Energy metabolism in unrestrained flatworms and annelids and factors controlling it]. Zh Obshch Biol 1974;35(6):874-885.

Kapu MM, Schaeffer DJ. Planarians in toxicology. Responses of asexual Dugesia dorotocephala to selected metals. Bull Environ Contam Toxicol 1991;47(2):302-307.

Kartry AL, Keith-Lee P, Morton WD. Planaria: Memory Transfer through Cannibalism Reexamined. Science 1964;146:274-275.

Kashikawa M, Agata K. [RNA-binding proteins in stem cell system of planarians]. Tanpakushitsu Kakusan Koso 2003;48(4 Suppl):438-443.

Kato C, Mihashi K, Ishida S. Motility recovery during the process of regeneration in freshwater planarians. Behav Brain Res 2004;150(1-2):9-14.

Kato K, Orii H, Watanabe K, Agata K. The role of dorsoventral interaction in the onset of planarian regeneration. Development 1999;126(5):1031-1040.

Kato K, Orii H, Watanabe K, Agata K. Dorsal and ventral positional cues required for the onset of planarian regeneration may reside in differentiated cells. Dev Biol 2001;233(1):109-121.

Kendall K, Nachtwey DS. Inhibition of regeneration of planaria by midlethal exposures to x rays. USNRDL-TR-68-29. Res Dev Tech Rep 1968:1-20.

Kenk R. Fresh-water triclads (Turbellaria) of North America. VII. The genus Macrocotyla. Trans Am Microsc Soc 1975;94(3):324-339.

Kimmel HD, Carlyon WD. Persistent effects of a serotonin depletor (p-chlorophenylalanine) in regenerated planaria (Dugesia dorotocephala). Behav Neurosci 1990;104(1):127-134.

Kimmel HD, Garrigan HA. Resistance to extinction in planaria. J Exp Psychol 1973;101(2):343-347.

Kitamura Y, Inden M, Sanada H, Takata K, Taniguchi T, Shimohama S, Orii H, Mochii M, Agata K, Watanabe K. Inhibitory effects of antiparkinsonian drugs and caspase inhibitors in a parkinsonian flatworm model. J Pharmacol Sci 2003;92(2):137-142.

Kitamura Y, Kakimura J, Taniguchi T. Protective effect of talipexole on MPTP-treated planarian, a unique parkinsonian worm model. Jpn J Pharmacol 1998;78(1):23-29.

Kliks MM, Palumbo NE. Eosinophilic meningitis beyond the Pacific Basin: the global dispersal of a peridomestic zoonosis caused by Angiostrongylus cantonensis, the nematode lungworm of rats. Soc Sci Med 1992;34(2):199-212.

Kobayashi C, Nogi T, Watanabe K, Agata K. Ectopic pharynxes arise by regional reorganization after anterior/posterior chimera in planarians. Mech Dev 1999;89(1-2):25-34.

Kobayashi C, Watanabe K, Agata K. The process of pharynx regeneration in planarians. Dev Biol 1999;211(1):27-38.

Kobayashi K, Arioka S, Hase S, Hoshi M. Signification of the sexualizing substance produced by the sexualized planarians. Zoolog Sci 2002;19(6):667-672.

Kobayashi K, Arioka S, Hoshi M. Seasonal changes in the sexualization of the planarian Dugesia ryukyuensis. Zoolog Sci 2002;19(11):1267-1278.

Kobayashi K, Hoshi M. Switching from asexual to sexual reproduction in the planarian Dugesia ryukyuensis: change of the fissiparous capacity along with the sexualizing process. Zoolog Sci 2002;19(6):661-666.

Kobayashi K, Matsumoto M, Hoshi M. [Switching mechanism from asexual to sexual reproduction in planarians]. Tanpakushitsu Kakusan Koso 2004;49(2):102-107.

Koinuma S, Umesono Y, Watanabe K, Agata K. Planaria FoxA (HNF3) homologue is specifically expressed in the pharynx-forming cells. Gene 2000;259(1-2):171-176.

Koinuma S, Umesono Y, Watanabe K, Agata K. The expression of planarian brain factor homologs, DjFoxG and DjFoxD. Gene Expr Patterns 2003;3(1):21-27.

Koopowitz H, Chien P. Ultrastructure of nerve plexus in flatworms. II. Sites of synaptic interactions. Cell Tissue Res 1975;157(2):207-216.

Kostelecky J, Elliott B, Schaeffer DJ. Planarians in toxicology. I. Physiology of sexual-only Dugesia dorotocephala: effects of diet and population density on adult weight and cocoon production. Ecotoxicol Environ Saf 1989;18(3):286-295.

Kotomin AV, Dontsova GV. [Changes in buoyant density of planarian mitochondria under starvation and nutrition]. Biokhimiia 1980;45(1):48-50.

Kouyoumjian HH, Villeneuve JP. Further studies on the toxicity of DDT to Planaria. Bull Environ Contam Toxicol 1979;22(1-2):108-112.

Kreshchenko ND, Sheiman IM, Fesenko EE. [Effect of weak electromagnetic radiation on regeneration of the pharynx in Dugesia tigrina planaria]. Ontogenez 2001;32(2):148-153.

Krichinskaia EB. [Cellular origins of regeneration in planarians. Current concepts of neoblasts]. Arkh Anat Gistol Embriol 1980;79(12):102-109.

Krichinskaia EB, Efimova GV. [Regeneration of a whole worm from a small fragment of the body of Dugesia tigrina planaria following repeated removal of regenerates]. Ontogenez 1978;9(5):510-514.

Krichinskaia EB, Korsakova L. [Effect of colchamine on regenerative processes in body fragments of Dugesia tigrina (Girard) planaria]. Nauchnye Doki Vyss Shkoly Biol Nauki 1981(9):64-68.

Krichinskaia EB, Malikova IG. [Reparation processes in asexual multiplication and regeneration of planaria Dugesia tigrina]. Arkh Anat Gistol Embriol 1969;57(11):48-51.

Krichinskaia EB, Serebrennik ME. [Effect of some pyrimidine derivatives on the regeneration of the planarian Dugesia tigrina]. Nauchnye Doki Vyss Shkoly Biol Nauki 1975(1):33-35.

Krichinskaya EB, Martynova MG. Distribution of neoblasts and mitoses during the asexual reproduction of the planarian Dugesia tigrina (Girard). Sov J Dev Biol 1975;5(4):309-314.

Krylov OA. [Reproduction of conditioned reflex activity by the administration of biochemical substrates]. Usp Fiziol Nauk 1974;5(4):22-51.

Krylov OA, Nazarian OA. [Reproduction of conditioned reflexes under the influence of homogenates in regenerating planaria]. Zh Vyssh Nerv Deiat Im I P Pavlova 1973;23(6):1303-1305.

Kumazaki T, Hori H, Osawa S. The nucleotide sequences of 5S rRNAs from two ribbon worms: Emplectonema gracile contains two 5S rRNA species differing considerably in their sequences. Nucleic Acids Res 1983;11(20):7141-7144.

Kurimoto K, Muto Y, Obayashi N, Terada T, Shirouzu M, Yabuki T, Aoki M, Seki E, Matsuda T, Kigawa T, Okumura H, Tanaka A, Shibata N, Kashikawa M, Agata K, Yokoyama S. Crystal structure of the N-terminal RecA-like domain of a DEAD-box RNA helicase, the Dugesia japonica vasa-like gene B protein. J Struct Biol 2005;150(1):58-68.

Kusayama T, Watanabe S. Reinforcing effects of methamphetamine in planarians. Neuroreport 2000;11(11):2511-2513.

Kuznedelov KD, Dziuba EV. [The determination of the species classification of Baikal planarian cocoons found in the stomach of the black grayling (Thymallus arcticus baicalensis) by a comparative analysis of the nucleotide sequences of the ribosomal RNA gene]. Zh Obshch Biol 1999;60(4):445-449.

Kuznedelov KD, Novikova OA, Naumova TV. [The molecular genetic typification of planarians in the genus Bdellocephala (Dendrocoelidae, Tricladida, Turbellaria) from Lake Baikal with an assessment of their species diversity]. Zh Obshch Biol 2000;61(3):336-344.

Kuznedelov KD, Timoshkin OA, Kumarev VP. [Molecular phylogeny of planarians (Turbellaria, Tricladida, Paludicola) from the lake Baikal. Use of comparative analysis for determination of nucleotide sequence of 18S ribosomal RNA]. Mol Biol (Mosk) 1996;30(6):1316-1325.

Lamatsch DK, Sharbel TF, Martin R, Bock C. A drop technique for flatworm chromosome preparation for light microscopy and high-resolution scanning electron microscopy. Chromosome Res 1998;6(8):654-656.

Landsperger WJ, Peters EH, Dresden MH. Properties of a collagenolytic enzyme from Bipalium kewense. Biochim Biophys Acta 1981;661(2):213-220.

Lange CS. Observations on some tumours found in two species of planaria--Dugesia etrusca and D. ilvana. J Embryol Exp Morphol 1966;15(2):125-130.

Lange CS. A quantitative study of the number and distribution of neoblasts in Dugesia lugubris (Planaria) with reference to size and ploidy. J Embryol Exp Morphol 1967;18(2):199-213.

Lange CS, Steele VE. The mechanism of anterior-posterior polarity control in planarians. Differentiation 1978;11(1):1-12.

Langer R, Fefferman M, Gryska P, Bergman K. A simple method for studying chemotaxis using sustained release of attractants from inert polymers. Can J Microbiol 1980;26(2):274-278.

Le moigne A. [Presence of nuclear emissions frequently associated with mitochondria in embryonic cells of Planaria]. C R Seances Soc Biol Fil 1967;161(3):508-511.

Le Moigne A. [Study of embryonic development and regeneration in the embryo of Planaria Polycelis nigra (Turbellarie, Triclade)]. J Embryol Exp Morphol 1966;15(1):39-60.

Le Moigne A, Martelly I. [Effects of the preventive or differed use of actinomycin D on the regeneration of young and adult Planaria]. C R Acad Sci Hebd Seances Acad Sci D 1974;279(4):367-369.

Lee RM. Conditioning of a free operant response in planaria. Science 1963;139:1048-1049.

Lender T. [Demonstration of the organizing role of the brain in regeneration of the eyes in planaria Polycelis nigra by grafting methods.]. C R Seances Soc Biol Fil 1950;144(19-20):1407-1409.

Lender T. [Regeneration of the eyes of the planaria Polycelis nigra in the presence of a mash of the anterior region of the body.]. C R Hebd Seances Acad Sci 1954;238(17):1742-1744.

Lender T. [Some properties of the organisine of eye regeneration in the planaria Polycelis nigra.]. C R Hebd Seances Acad Sci 1955;240(17):1726-1728.

Lender T. [The specific inhibition of the differentiation of the brain of fresh water planarians during regeneration.]. J Embryol Exp Morphol 1960;8:291-301.

Lender T. [Role of neurosecretion during regeneration and asexual reproduction of freshwater planaria]. Ann Endocrinol (Paris) 1970;31(3):463-466.

Lender T, Gabriel A. [Neoblasts Labelled with Tritiated Uridine Migrate and Construct the Regeneration Blastema in Fresh-Water Planaria.]. C R Hebd Seances Acad Sci 1965;260:4095-4097.

Lender T, Zghal F. [Influence of the brain and neurosecretion on scissiparity of the planaria Dugesia gonocephala]. C R Acad Sci Hebd Seances Acad Sci D 1968;267(23):2008-2009.

Lenicque PM. [Control of the regeneration of small pieces of the planarian Dugesia tigrina by cylic AMP and GMP nucleotides]. C R Acad Sci Hebd Seances Acad Sci D 1976;283(11 D):1317-1319.

Lentati GB. On the relationship between the chromosome set and the sexual differentiation of the neoblasts in planarians. Acta Embryol Exp (Palermo) 1976(3):299-317.

Lentz TL. Rhabdite formation in planaria: the role of microtubules. J Ultrastruct Res 1967;17(1):114-126.

Lints TJ, Parsons LM, Hartley L, Lyons I, Harvey RP. Nkx-2.5: a novel murine homeobox gene expressed in early heart progenitor cells and their myogenic descendants. Development 1993;119(2):419-431.

Liotti FS. [Studies and investigations of regeneration in Planaria. Inhibiting action of amphetamine and other sympathicominetic amines on the regeneration of the head in Dugesia lugubris.]. Riv Biol 1961;54:415-455.

Liotti FS. [Relations between the Chemical Structure of Some Sympathomimetic and Sympatholytic Drugs and Their Capacity to Inhibit the Regeneration of the Head in Planaria.]. Boll Soc Ital Biol Sper 1963;39:921-925.

Liotti FS, Andreoli V. [Studies and Research on Regeneration in Planaria. Further Observations on the Effect of Amphetamine on Regeneration of the Head in Dugesia Lugubris.]. Arch Sci Biol (Bologna) 1964;48:1-17.

Liotti FS, Bruschelli G. Relations between the Nervous System and Regeneration in the Planarians. First Observations on the Regenerative Capacity of Specimens of Dugesia Lugubris before and after the Opening of the Cocoons. Riv Biol 1964;57:SUPPL 2:137-149.

Liotti FS, Maitino L. [The problem of supernumerary eyes in the light of current knowledge of regenerative processes in Planaria]. Riv Biol 1968;61(4):349-370.

Lord BA, DiBona DR. Role of the septate junction in the regulation of paracellular transepithelial flow. J Cell Biol 1976;71(3):967-972.

Lozeron H, Maggiora A, Jadassohn W. [Experimental studies of griseofulvin. Experiments on guinea pigs, Vicia faba equina, Planaria gonocephala and Paracentrotus lividus]. G Ital Dermatol Minerva Dermatol 1966;107(5):799-808.

Luk'ianov KA, Tarabykin VS, Potapov VK, Bekman EP, Luk'ianov SA. [The cloning of fragments of homeobox genes expressed during planarian regeneration]. Ontogenez 1994;25(6):28-32.

Lytle CD, Hsu J, Faw JM, Miller CW. Responses of the planarian Dugesia dorotocephala to ultraviolet radiation or photosensitized treatments. Photochem Photobiol 1988;48(4):457-459.

Macan TT. The influence of predation on the composition of fresh-water animal communities. Biol Rev Camb Philos Soc 1977;52(1):45-70.

Macrae EK. Localization of porphyrin fluorescence in planarians. Science 1961;134:331-332.

Malczewska M, Czubaj A, Morawska E, Moraczewski J. Changes in the acetylcholinesterase activity during the regeneration of planarian Dugesia lugubris (O. Schmidt) ultrastructural studies. Acta Med Pol 1980;21(4):381-382.

Mannini L, Rossi L, Deri P, Gremigni V, Salvetti A, Salo E, Batistoni R. Djeyes absent (Djeya) controls prototypic planarian eye regeneration by cooperating with the transcription factor Djsix-1. Dev Biol 2004;269(2):346-359.

Margotta V, Caronti B, Meco G, Merante A, Ruggieri S, Venturini G, Palladini G. Effects of cocaine treatment on the nervous system of planaria (Dugesia gonocephala s. l.). Histochemical and ultrastructural observations. Eur J Histochem 1997;41(3):223-230.

Margotta V, Hernandez MC, Cappiello-Valcamonica A, Palladini G, Albani LM. Action of several food colours on whole and regenerant Planaria. Riv Biol 1979;72(1-2):11-38.

Marinelli M, Farnesi RM. [The cement glands in Dugesia lugubris: histochemical and ultrastructural studies]. Riv Biol 1974;67(3):301-317.

Marsal M, Pineda D, Salo E. Gtwnt-5 a member of the wnt family expressed in a subpopulation of the nervous system of the planarian Girardia tigrina. Gene Expr Patterns 2003;3(4):489-495.

Martelly I. [Variations in DNA synthesis and the activities of acid and alkaline deoxyribonucleases during regeneration of planaria (Polycelis tenuis)]. C R Seances Acad Sci D 1980;290(24):1571-1574.

Martelly I. Calcium thresholds in the activation of DNA and RNA synthesis in cultured planarian cells: relationship with hormonal and DB cAMP effects. Cell Differ 1984;15(1):25-36.

Martelly I, Borney C, Moigne AL. [Demonstration of 2 phases in RNA synthesis during regeneration of planarian (Polycelis tenuis-nigra)]. C R Acad Sci Hebd Seances Acad Sci D 1976;282(20):1805-1808.

Martelly I, Le Moigne A. Ribonucleic acid metabolism during planarian regeneration. Reprod Nutr Dev 1980;20(5A):1527-1537.

Martelly I, Le Moigne A. [Comparison of the effects of actinomycin D on RNA synthesis of young and adult regenerating planarians]. C R Seances Soc Biol Fil 1980;173(6):1023-1030.

Martelly I, Molla A, Thomasset M, Le Moigne A. Planarian regeneration: in vivo and in vitro effects of calcium and calmodulin on DNA synthesis. Cell Differ 1983;13(1):25-34.

Marvillet C. [Anatomic variability and distribution of the troglobe planaria Dendrocoelum (Dendrocoelides) collini of Beauchamp]. C R Acad Sci Hebd Seances Acad Sci D 1967;264(21):2512-2515.

Mason PR. Chemo-klino-kinesis in planarian food location. Anim Behav 1975;23(2):460-469.

Mats MV, Shagin DA, Usman N, Bogdanova EA, Fradkov AF, Soboleva TA, Luk'ianov SA. [Cloning of region-specific genetic markers of planaria using a new method--ordered differential display]. Bioorg Khim 1998;24(12):910-915.

Matsumoto M, Kobayashi K, Hoshi M. [Origin of germ cells in planarians]. Tanpakushitsu Kakusan Koso 2005;50(6 Suppl):541-545.

Matz M, Usman N, Shagin D, Bogdanova E, Lukyanov S. Ordered differential display: a simple method for systematic comparison of gene expression profiles. Nucleic Acids Res 1997;25(12):2541-2542.

McConnell JV, Cornwell PR, Clay M. An apparatus for conditioning Planaria. Am J Psychol 1960;73:618-622.

McKanna JA. Fine structure of the protonephridial system in Planaria. II. Ductules, collecting ducts, and osmoregulatory cells. Z Zellforsch Mikrosk Anat 1968;92(4):524-535.

McKanna JA. Fine structure of the protonephridial system in Planaria. I. Flame cells. Z Zellforsch Mikrosk Anat 1968;92(4):509-523.

Medvedev IV, Gremiachikh VA, Zheltov SV, Bogdanenko OV, Aksenova IA. [Regeneration of photoreceptor organs in freshwater planarians at different levels of accumulation of natural methylmercury compounds]. Ontogenez 2006;37(2):136-141.

Medvedev IV, Komov VT. [Regeneration of freshwater planarians Dugesia tigrina and Polycelis tenuis under the influence of methyl mercury compounds of natural origin]. Ontogenez 2005;36(1):35-40.

Melo AS, Andrade CF. Differential predation of the planarian Dugesia tigrina on two mosquito species under laboratory conditions. J Am Mosq Control Assoc 2001;17(1):81-83.

Michetti F, Cocchia D. S-100-like immunoreactivity in a planarian. An immunochemical and immunocytochemical study. Cell Tissue Res 1982;223(3):575-582.

Minelli A. Microplana mahnerti n. sp., a new terrestrial planarian from Jura mts. (Turbellaria Tricladida Rhynchodemidae). Rev Suisse Zool 1977;84(1):173-176.

Mineta K, Nakazawa M, Cebria F, Ikeo K, Agata K, Gojobori T. Origin and evolutionary process of the CNS elucidated by comparative genomics analysis of planarian ESTs. Proc Natl Acad Sci U S A 2003;100(13):7666-7671.

Mitchell SR, Beaton JM, Bradley RJ. Biochemical transfer of acquired information. Int Rev Neurobiol 1975;17:61-83.

Mitra S, Kar S, Aditya AK. Impact of CdCl2 on biochemical changes in planaria, Dugesia bengalensis Kawakatsu. Indian J Exp Biol 2003;41(8):921-923.

Mochizuki K, Nishimiya-Fujisawa C, Fujisawa T. Universal occurrence of the vasa-related genes among metazoans and their germline expression in Hydra. Dev Genes Evol 2001;211(6):299-308.

Monkiedje A, Njinel T, Meyabeme Elono AL, Zebaze SH, Kemka N, Tchounwou PB, Djomo JE. Freshwater microcosms-based assessment of eco-toxicological effects of a chemical effluent from the Pilcam industry in Cameroon. Int J Environ Res Public Health 2004;1(2):111-123.

Moraczewski J, Duma A. Localization of adenylate cyclase activity in the tissues of an intact planarian Dugesia lugubris (O. Schmidt). Histochemistry 1981;71(2):301-311.

Morita M. Electron microscopic studies on Planaria. I. Fine structure of muscle fiber in the head of the planarian Dugesia dorotocephala. J Ultrastruct Res 1965;13(5):383-395.

Morita M. Electron microscopic studies on planaria. IV. Fine structure of some secretory gland in the planarian Dugesia dorotocephala. Fukushima J Med Sci 1968;15(1):13-33.

Morita M, Best JB. Electron microscopic studies on Planaria. II. Fine structure of the neurosecretory system in the planarian Dugesia dorotocephala. J Ultrastruct Res 1965;13(5):396-408.

Morita M, Best JB. Electron microscopic studies of Planaria. 3. Some observations on the fine structure of planarian nervous tissue. J Exp Zool 1966;161(3):391-411.

Morita M, Hall F, Best JB, Gern W. Photoperiodic modulation of cephalic melatonin in planarians. J Exp Zool 1987;241(3):383-388.

Munoz-Marmol AM, Casali A, Miralles A, Bueno D, Bayascas JR, Romero R, Salo E. Characterization of platyhelminth POU domain genes: ubiquitous and specific anterior nerve cell expression of different epitopes of GtPOU-1. Mech Dev 1998;76(1-2):127-140.

Nagata Y. [Memory substance (author's transl)]. No Shinkei Geka 1974;2(1):23-27.

Nakazawa M, Cebria F, Mineta K, Ikeo K, Agata K, Gojobori T. Search for the evolutionary origin of a brain: planarian brain characterized by microarray. Mol Biol Evol 2003;20(5):784-791.

Nano GM, Binello A, Bianco MA, Ugazio G, Burdino S. In vitro tests to evaluate potential biological activity in natural substances. Fitoterapia 2002;73(2):140-146.

Nazarian OA. [Preservation of traces from a previous experience in planaria following regeneration]. Zh Vyssh Nerv Deiat Im I P Pavlova 1973;23(1):120-125.

Nazarian OA. [The effect of a steady magnetic field on the permanence of conditioned reflex elaboration in planaria]. Zh Vyssh Nerv Deiat Im I P Pavlova 1974;24(1):183-185.

Nelson FR, Edmond M, Mohamed AK. Effects of selected insect growth regulators and pesticides on Dugesia dorotocephala and Dugesia tigrina (Tricladida: Turbellaria). J Am Mosq Control Assoc 1988;4(2):184-186.

Nelson FR, Gray J, Aikhionbare F. Tolerance of the planarian Dugesia tigrina (Tricladida: Turbellaria) to pesticides and insect growth regulators in a small-scale field study. J Am Mosq Control Assoc 1994;10(1):104-105.

Nelson FR, Holloway D, Mohamed AK. A laboratory study of cyromazine on Aedes aegypti and Culex quinquefasciatus and its activity on selected predators of mosquito larvae. J Am Mosq Control Assoc 1986;2(3):296-299.

Nentwig MR. Comparative morphological studies of head development after decapitation and after fission in the planarian Dugesia dorotocephala. Trans Am Microsc Soc 1978;97(3):297-310.

Nentwig MR. A morphological study of the effects of colcemid on head regeneration in Dugesia dorotocephala. Acta Embryol Exp (Palermo) 1978(1):113-129.

Ness DK, Foley GL, Villar D, Hansen LG. Effects of 3-iodo-L-tyrosine, a tyrosine hydroxylase inhibitor, on eye pigmentation and biogenic amines in the planarian, Dugesia dorotocephala. Fundam Appl Toxicol 1996;30(2):153-161.

Newmark PA. Opening a new can of worms: a large-scale RNAi screen in planarians. Dev Cell 2005;8(5):623-624.

Newmark PA, Reddien PW, Cebria F, Sanchez Alvarado A. Ingestion of bacterially expressed double-stranded RNA inhibits gene expression in planarians. Proc Natl Acad Sci U S A 2003;100 Suppl 1:11861-11865.

Newmark PA, Sanchez Alvarado A. Bromodeoxyuridine specifically labels the regenerative stem cells of planarians. Dev Biol 2000;220(2):142-153.

Newmark PA, Sanchez Alvarado A. Not your father's planarian: a classic model enters the era of functional genomics. Nat Rev Genet 2002;3(3):210-219.

Nogi T, Levin M. Characterization of innexin gene expression and functional roles of gap-junctional communication in planarian regeneration. Dev Biol 2005;287(2):314-335.

Nogi T, Watanabe K. Position-specific and non-colinear expression of the planarian posterior (Abdominal-B-like) gene. Dev Growth Differ 2001;43(2):177-184.

Nogi T, Yuan YE, Sorocco D, Perez-Tomas R, Levin M. Eye regeneration assay reveals an invariant functional left-right asymmetry in the early bilaterian, Dugesia japonica. Laterality 2005;10(3):193-205.

Novikov VV, Sheiman IM, Fesenko EE. [Effect of weak and superweak magnetic fields on intensity and asexual reproduction of the planarian Dugesia tigrina]. Biofizika 2002;47(1):125-129.

Novikov VV, Sheiman IM, Lisitsyn AS, Kliubin AV, Fesenko EE. [Dependence of effects of weak combined low-frequency variable and constant magnetic fields on the intensity of asexual reproduction of planarians Dugesia tigrina on the magnitude of the variable field]. Biofizika 2002;47(3):564-567.

Ogawa K, Ishihara S, Saito Y, Mineta K, Nakazawa M, Ikeo K, Gojobori T, Watanabe K, Agata K. Induction of a noggin-like gene by ectopic DV interaction during planarian regeneration. Dev Biol 2002;250(1):59-70.

Ogawa K, Kobayashi C, Hayashi T, Orii H, Watanabe K, Agata K. Planarian fibroblast growth factor receptor homologs expressed in stem cells and cephalic ganglions. Dev Growth Differ 2002;44(3):191-204.

Ogawa K, Wakayama A, Kunisada T, Orii H, Watanabe K, Agata K. Identification of a receptor tyrosine kinase involved in germ cell differentiation in planarians. Biochem Biophys Res Commun 1998;248(1):204-209.

Ohama T, Kumazaki T, Hori H, Osawa S, Takai M. Fresh-water planarias and a marine planaria are relatively dissimilar in the 5S rRNA sequences. Nucleic Acids Res 1983;11(2):473-476.

Okamoto K, Takeuchi K, Agata K. Neural projections in planarian brain revealed by fluorescent dye tracing. Zoolog Sci 2005;22(5):535-546.

Oliver G, Vispo M, Mailhos A, Martinez C, Sosa-Pineda B, Fielitz W, Ehrlich R. Homeoboxes in flatworms. Gene 1992;121(2):337-342.

Onwumere EA, Wells MR. Metabolism of p,p'-DDT by the freshwater Planarian Phagocata gracilis. Bull Environ Contam Toxicol 1983;31(1):18-21.

Orii H, Agata K, Watanabe K. POU-domain genes in planarian Dugesia japonica: the structure and expression. Biochem Biophys Res Commun 1993;192(3):1395-1402.

Orii H, Ito H, Watanabe K. Anatomy of the planarian Dugesia japonica I. The muscular system revealed by antisera against myosin heavy chains. Zoolog Sci 2002;19(10):1123-1131.

Orii H, Kato K, Umesono Y, Sakurai T, Agata K, Watanabe K. The planarian HOM/ HOX homeobox genes (Plox) expressed along the anteroposterior axis. Dev Biol 1999;210(2):456-468.

Orii H, Mochii M, Watanabe K. A simple "soaking method" for RNA interference in the planarian Dugesia japonica. Dev Genes Evol 2003;213(3):138-141.

Orii H, Sakurai T, Watanabe K. Distribution of the stem cells (neoblasts) in the planarian Dugesia japonica. Dev Genes Evol 2005;215(3):143-157.

Ortabaeva LM. [Dynamics of succinate dehydrogenase activity in neoblasts during regeneration of Dendrocoelum lacteum planaria]. Tsitologiia 1973;15(12):1476-1480.

Osawa S, Jukes TH, Watanabe K, Muto A. Recent evidence for evolution of the genetic code. Microbiol Rev 1992;56(1):229-264.

Oviedo NJ, Newmark PA, Sanchez Alvarado A. Allometric scaling and proportion regulation in the freshwater planarian Schmidtea mediterranea. Dev Dyn 2003;226(2):326-333.

Ozaki K, Hara R, Hara T. Histochemical localization of retinochrome and rhodopsin studied by fluorescence microscopy. Cell Tissue Res 1983;233(2):335-345.

Palakodeti D, Smielewska M, Graveley BR. MicroRNAs from the Planarian Schmidtea mediterranea: A model system for stem cell biology. Rna 2006.

Palladini G, Hernandez MC, Margotta V, Cappiello-Valcamonica A, Carolei A. Effects of different anion solutions of varying osmotic pressure and viscosity on the dopaminergic nervous system of Dugesia gonocephala S.L. Riv Biol 1978;71(1-4):41-61.

Palladini G, Margotta V, Carolei A, Chiarini F, Del Piano M, Lauro GM, Medolago-Albani L, Venturini G. The cerebrum of Dugesia gonocephala s.l. Platyhelminthes, Turbellaria, Tricladida. Morphological and functional observations. J Hirnforsch 1983;24(2):165-172.

Palladini G, Margotta V, Carolei A, Conforti A. [Pharmacological, electronmicroscopic and histochemical study on an in-vivo model of the dopaminergic nervous system (Dugesia gonocephala s.l)]. Acta Neurol (Napoli) 1976;31(1):1-5.

Palladini G, Margotta V, Carolei A, Hernandez MC. Dopamine agonist performance in Planaria after manganese treatment. Experientia 1980;36(4):449-450.

Palladini G, Margotta V, Medolago-Albani L, Hernandez MC, Carolei A, de Michele T. Photosensitization and the nervous system in the planarian Dugesia gonocephala. A histochemical, ultrastructural and behavioral investigation. Cell Tissue Res 1981;215(2):271-279.

Palladini G, Medolago-Albani L, Gallo VP, Lauro GM, Diana G, Scorsini D, Alfei L, Margotta V. The answer of the planarian Dugesia gonocephala neurons to nerve growth factor. Cell Mol Biol 1988;34(1):53-63.

Palladini G, Medolago-Albani L, Margotta V, Conforti A, Carolei A. The pigmentary system of planaria. I. Morphology. Cell Tissue Res 1979;199(2):197-202.

Palladini G, Medolago-Albani L, Margotta V, Conforti A, Carolei A. The pigmentary system of planaria. II. Physiology and functional morphology. Cell Tissue Res 1979;199(2):203-211.

Palladini G, Ruggeri S, Stocchi F, De Pandis MF, Venturini G, Margotta V. A pharmacological study of cocaine activity in planaria. Comp Biochem Physiol C Pharmacol Toxicol Endocrinol 1996;115(1):41-45.

Paris M, Lenicque PM. [Effect of tetrahydrocannabinol and of cannabidiol on wound healing and regeneration of the planaria Dugesia tigrina]. Therapie 1975;30(1):97-102.

Pascoline R. [Research on the inhibition of regeneration in bicephalic planaria]. Boll Soc Ital Biol Sper 1965;41(17):1022-1024.

Pascolini R. [Research on the regeneration of the pharynx in Planaria]. Boll Soc Ital Biol Sper 1966;42(15):968-969.

Pascolini R. [Inhibitory action of Rana esculenta cerebral poltiglia on the development of the cerebral ganglia of Planaria]. Boll Soc Ital Biol Sper 1966;42(3):123-124.

Pascolini R, Di Rosa I, Fagotti A, Panara F, Gabbiani G. The mammalian anti-alpha-smooth muscle actin monoclonal antibody recognizes an alpha-actin-like protein in planaria (Dugesia lugubris s.l.). Differentiation 1992;51(3):177-186.

Pascolini R, Gargiulo AM, Spreca A, Orlacchio A. Studies on the cytochemical localization of adenylate-cyclase activity in Dugesia lugubris s.I. Experientia 1979;35(10):1315-1317.

Pascolini R, Lorvik S, Maci R, Camatini M. Immunoelectron microscopic localization of actin in migrating cells during planarian wound healing. Tissue Cell 1988;20(2):157-163.

Pascolini R, Panara F, Di Rosa I, Fagotti A, Lorvik S. Characterization and fine-structural localization of actin- and fibronectin-like proteins in planaria (Dugesia lugubris s.l.). Cell Tissue Res 1992;267(3):499-506.

Pascolini R, Tei S, Vagnetti D, Bondi C. Epidermal cell migration during wound healing in Dugesia lugubris. Observations based on scanning electron microscopy and treatment with cytochalasin. Cell Tissue Res 1984;236(2):345-349.

Passarelli F, Merante A, Pontieri FE, Margotta V, Venturini G, Palladini G. Opioid-dopamine interaction in planaria: a behavioral study. Comp Biochem Physiol C Pharmacol Toxicol Endocrinol 1999;124(1):51-55.

Patra BC, Aditya AK. Circular form of regeneration in an unidentified species of land planarians, Bipalium sp. Indian J Exp Biol 2001;39(5):496-499.

Pavlova GA. Ciliary locomotion of the mollusk is different from that of the planaria. Dokl Biol Sci 2000;375:630-632.

Pearson H. The regeneration gap. Nature 2001;414(6862):388-390.

Pedersen KJ. Slime-secreting cells of planarians. Ann N Y Acad Sci 1963;106:424-443.

Pennisi E. Evolution and development. Comparative biology joins the molecular age. Science 2002;296(5574):1792-1795.

Pennisi E. Evolution of developmental diversity. Evo-devo devotees eye ocular origins and more. Science 2002;296(5570):1010-1011.

Pennisi E. Evolution of developmental diversity meeting. RNAi takes Evo-Devo world by storm. Science 2004;304(5669):384.

Perekalin DV. [DNA reproduction in the organism of planarians in post-injury recovery]. Tsitologiia 1975;17(9):1084-1086.

Perich MJ, Clair PM, Boobar LR. Integrated use of planaria (Dugesia dorotocephala) and Bacillus thuringiensis var. israelensis against Aedes taeniorhynchus: a laboratory bioassay. J Am Mosq Control Assoc 1990;6(4):667-671.

Pescetto G, Dettore D. [Negative phototaxis and conditioning in the planarian Dugesia dorotocephala]. Riv Neurobiol 1982;27(2):287-295.

Peter R, Wolfrum DI, Neuhoff V. Micro-electrophoresis in continuous polyacrylamide gradient gels for the analytical separation of protein extracts from planarians, (Platyhelminthes: Turbellaria tricladida). Comp Biochem Physiol B 1976;55(4):583-589.

Phillips J, Wells M, Chandler C. Metabolism of DDT by the freshwater planarian, Phagocata velata. Bull Environ Contam Toxicol 1974;12(3):355-358.

Pigon A, Morita M, Best JB. Cephalic mechanism for social control of fissioning in planarians. II. Localization and identification of the receptors by electron micrographic and ablation studies. J Neurobiol 1974;5(5):443-462.

Pineda D, Gonzalez J, Callaerts P, Ikeo K, Gehring WJ, Salo E. Searching for the prototypic eye genetic network: Sine oculis is essential for eye regeneration in planarians. Proc Natl Acad Sci U S A 2000;97(9):4525-4529.

Pineda D, Rossi L, Batistoni R, Salvetti A, Marsal M, Gremigni V, Falleni A, Gonzalez-Linares J, Deri P, Salo E. The genetic network of prototypic planarian eye regeneration is Pax6 independent. Development 2002;129(6):1423-1434.

Pineda D, Salo E. Planarian Gtsix3, a member of the Six/so gene family, is expressed in brain branches but not in eye cells. Gene Expr Patterns 2002;2(1-2):169-173.

Pongratz N, Gerace L, Michiels NK. Genetic differentiation within and between populations of a hermaphroditic freshwater planarian. Heredity 2002;89(1):64-69.

Pongratz N, Michiels NK. High multiple paternity and low last-male sperm precedence in a hermaphroditic planarian flatworm: consequences for reciprocity patterns. Mol Ecol 2003;12(6):1425-1433.

Pongratz N, Storhas M, Carranza S, Michiels NK. Phylogeography of competing sexual and parthenogenetic forms of a freshwater flatworm: patterns and explanations. BMC Evol Biol 2003;3:23.

Post S, Wells MR. Metabolism of aldrin by the freshwater planarian Phagocata gracilis. Bull Environ Contam Toxicol 1985;34(6):871-875.

Pra D, Lau AH, Knakievicz T, Carneiro FR, Erdtmann B. Environmental genotoxicity assessment of an urban stream using freshwater planarians. Mutat Res 2005;585(1-2):79-85.

Pronko NH, Wehrenberg L, Jr. A maze for planaria. Am J Psychol 1957;70(1):128.

Quick DC, Johnson RG. Gap junctions and rhombic particle arrays in planaria. J Ultrastruct Res 1977;60(3):348-361.

Raffa RB, Baron DA, Tallarida RJ. Schild (apparent pA2) analysis of a kappa-opioid antagonist in Planaria. Eur J Pharmacol 2006;540(1-3):200-201.

Raffa RB, Dasrath CS, Brown DR. Disruption of a drug-induced choice behavior by UV light. Behav Pharmacol 2003;14(7):569-571.

Raffa RB, Desai P. Description and quantification of cocaine withdrawal signs in Planaria. Brain Res 2005;1032(1-2):200-202.

Raffa RB, Holland LJ, Schulingkamp RJ. Quantitative assessment of dopamine D2 antagonist activity using invertebrate (Planaria) locomotion as a functional endpoint. J Pharmacol Toxicol Methods 2001;45(3):223-226.

Raffa RB, Martley AF. Amphetamine-induced increase in planarian locomotor activity and block by UV light. Brain Res 2005;1031(1):138-140.

Raffa RB, Stagliano GW, Umeda S. kappa-Opioid withdrawal in Planaria. Neurosci Lett 2003;349(3):139-142.

Raffa RB, Valdez JM. Cocaine withdrawal in Planaria. Eur J Pharmacol 2001;430(1):143-145.

Raffa RB, Valdez JM, Holland LJ, Schulingkamp RJ. Energy-dependent UV light-induced disruption of (-)sulpiride antagonism of dopamine. Eur J Pharmacol 2000;406(3):R11-12.

Ramachandran S, Beukeboom LW, Gerace L, Pavlovic N, Carranza S, Michiels NK. Isolation and characterization of microsatellite loci from the planarian Dugesia polychroa (Schmidt) (Platyhelminthes:Tricladida). Mol Ecol 1997;6(4):389-391.

Rand HW, Browne A. Inhibition of Regeneration in Planarians by Grafting: Technique of Grafting. Proc Natl Acad Sci U S A 1926;12(9):575-581.

Rand HW, Ellis M. Inhibition of Regeneration in Two-Headed or Two-Tailed Planarians. Proc Natl Acad Sci U S A 1926;12(9):570-574.

Ratner SC, Vandeventer JM. Effects of Water Current on Responses of Planaria to Light. J Comp Physiol Psychol 1965;60:138-139.

Rawls SM, Gomez T, Stagliano GW, Raffa RB. Measurement of glutamate and aspartate in Planaria. J Pharmacol Toxicol Methods 2006;53(3):291-295.

Rawls SM, Rodriguez T, Baron DA, Raffa RB. A nitric oxide synthase inhibitor (l-NAME) attenuates abstinence-induced withdrawal from both cocaine and a cannabinoid agonist (WIN 55212-2) in Planaria. Brain Res 2006;1099(1):82-87.

Rebrikov DV, Bogdanova EA, Bulina ME, Luk'ianov SA. [A new planarian extrachromosomal virus-like element revealed by subtraction hybridization]. Mol Biol (Mosk) 2002;36(6):1002-1011.

Rebrikov DV, Bulina ME, Bogdanova EA, Vagner LL, Lukyanov SA. Complete genome sequence of a novel extrachromosomal virus-like element identified in planarian Girardia tigrina. BMC Genomics 2002;3(1):15.

Reddien PW, Bermange AL, Murfitt KJ, Jennings JR, Sanchez Alvarado A. Identification of genes needed for regeneration, stem cell function, and tissue homeostasis by systematic gene perturbation in planaria. Dev Cell 2005;8(5):635-649.

Reddien PW, Oviedo NJ, Jennings JR, Jenkin JC, Sanchez Alvarado A. SMEDWI-2 is a PIWI-like protein that regulates planarian stem cells. Science 2005;310(5752):1327-1330.

Reddien PW, Sanchez Alvarado A. Fundamentals of planarian regeneration. Annu Rev Cell Dev Biol 2004;20:725-757.

Reddy A, Frazer BA, Fried B. Low molecular weight hydrophilic chemicals that attract Echinostoma trivolvis and E. caproni cercariae. Int J Parasitol 1997;27(3):283-287.

Redi CA, Garagna S, Pellicciari C. Chromosome preparation from planarian blastemas: a procedure suitable for cytogenetic and cytochemical studies. Stain Technol 1982;57(3):190-192.

Reynierse JH. Some effects of light on the formation of aggregations in planaria Phagocata gracilis. Anim Behav 1966;14(2):246-250.

Reynierse JH. Reactions to light in four species of planaria. J Comp Physiol Psychol 1967;63(2):366-368.

Reynierse JH. Aggregation formation in planaria, Phagocata gracilis and Cura foremani: species differentiation. Anim Behav 1967;15(2):270-272.

Reynierse JH, Ellis RR. Aggregation formation in three species of planaria: distance to nearest neighbour. Nature 1967;214(91):895-896.

Rice GE, Jr., Lawless RH. Behavior variability and reactive inhibition in the maze behavior of Planaria dorotocephala. J Comp Physiol Psychol 1957;50(1):105-108.

Robb SM, Sanchez Alvarado A. Identification of immunological reagents for use in the study of freshwater planarians by means of whole-mount immunofluorescence and confocal microscopy. Genesis 2002;32(4):293-298.

Rodrigo AG. Calibrating the bootstrap test of monophyly. Int J Parasitol 1993;23(4):507-514.

Rodriguez LV, Flickinger RA. Bipolar head regeneration in Planaria induced by chick embryo extracts. Biol Bull 1971;140(1):117-124.

Rohlich P, Torok LJ. [Electron microscopic studies on the eye of Planaria.]. Z Zellforsch Mikrosk Anat 1961;54:362-381.

Romero R, Bueno D. Disto-proximal regional determination and intercalary regeneration in planarians, revealed by retinoic acid induced disruption of regeneration. Int J Dev Biol 2001;45(4):669-673.

Rossi L, Deri P, Andreoli I, Gremigni V, Salvetti A, Batistoni R. Expression of DjXnp, a novel member of the SNF2-like ATP-dependent chromatin remodelling genes, in intact and regenerating planarians. Int J Dev Biol 2003;47(4):293-298.

Rossi L, Salvetti A, Lena A, Batistoni R, Deri P, Pugliesi C, Loreti E, Gremigni V. DjPiwi-1, a member of the PAZ-Piwi gene family, defines a subpopulation of planarian stem cells. Dev Genes Evol 2006;216(6):335-346.

Rulon O. The control of reconstitutional development in planarians with pilocarpine. Physiol Zool 1951;24(1):76-85.

Ryder TA, Bowen ID. The use of x-ray microanalysis to investigate problems encountered in enzyme cytochemistry. J Microsc 1974;101 Pt 2:143-151.

Sabourin TD, Faulk RT, Goss LB. The efficacy of three non-mammalian test systems in the identification of chemical teratogens. J Appl Toxicol 1985;5(4):227-233.

Sadauskas KK, Shuranova Zh P. [Method of recording the actograms of small aquatic animals and primary automatic processing of the information]. Zh Vyssh Nerv Deiat Im I P Pavlova 1982;32(6):1176-1179.

Saito Y, Koinuma S, Watanabe K, Agata K. Mediolateral intercalation in planarians revealed by grafting experiments. Dev Dyn 2003;226(2):334-340.

Saitoh O, Oshima T, Agata K, Watanabe K, Nakata H. Molecular cloning of a novel ADP-ribosylation factor (ARF) expressed in planarians. Biochim Biophys Acta 1996;1309(3):205-210.

Saitoh O, Oshima T, Agata K, Watanabe K, Nakata H. Corrigendum to "Molecular cloning of a novel ADP-ribosylation factor (ARF) expressed in planarians" [Biochim. Biophys. Acta 1309 (1996) 205-210]. Biochim Biophys Acta 1997;1352(2):231.

Saitoh O, Yuruzume E, Nakata H. Identification of planarian serotonin receptor by ligand binding and PCR studies. Neuroreport 1996;8(1):173-178.

Saitoh O, Yuruzume E, Watanabe K, Nakata H. Molecular identification of a G protein-coupled receptor family which is expressed in planarians. Gene 1997;195(1):55-61.

Sakai T, Kato K, Watanabe K, Orii H. Planarian pharynx regeneration revealed by the expression of myosin heavy chain-A. Int J Dev Biol 2002;46(3):329-332.

Sakharova N, Popkova LG, Pakhniushchaia NE. [Some regulatory mechanisms of asexual reproduction in planaria]. Ontogenez 1975;6(6):579-584.

Salo E. The power of regeneration and the stem-cell kingdom: freshwater planarians (Platyhelminthes). Bioessays 2006;28(5):546-559.

Salo E, Baguna J. Regeneration and pattern formation in planarians. I. The pattern of mitosis in anterior and posterior regeneration in Dugesia (G) tigrina, and a new proposal for blastema formation. J Embryol Exp Morphol 1984;83:63-80.

Salo E, Baguna J. Cell movement in intact and regenerating planarians. Quantitation using chromosomal, nuclear and cytoplasmic markers. J Embryol Exp Morphol 1985;89:57-70.

Salo E, Baguna J. Stimulation of cellular proliferation and differentiation in the intact and regenerating planarian Dugesia(G) tigrina by the neuropeptide substance P. J Exp Zool 1986;237(1):129-135.

Salo E, Baguna J. Regeneration in planarians and other worms: New findings, new tools, and new perspectives. J Exp Zool 2002;292(6):528-539.

Salvetti A, Batistoni R, Deri P, Rossi L, Sommerville J. Expression of DjY1, a protein containing a cold shock domain and RG repeat motifs, is targeted to sites of regeneration in planarians. Dev Biol 1998;201(2):217-229.

Salvetti A, Lena A, Rossi L, Deri P, Cecchettini A, Batistoni R, Gremigni V. Characterization of DeY1, a novel Y-box gene specifically expressed in differentiating male germ cells of planarians. Gene Expr Patterns 2002;2(3-4):195-200.

Salvetti A, Rossi L, Deri P, Batistoni R. An MCM2-related gene is expressed in proliferating cells of intact and regenerating planarians. Dev Dyn 2000;218(4):603-614.

Salvetti A, Rossi L, Lena A, Batistoni R, Deri P, Rainaldi G, Locci MT, Evangelista M, Gremigni V. DjPum, a homologue of Drosophila Pumilio, is essential to planarian stem cell maintenance. Development 2005;132(8):1863-1874.

Sanchez Alvarado A. The freshwater planarian Schmidtea mediterranea: embryogenesis, stem cells and regeneration. Curr Opin Genet Dev 2003;13(4):438-444.

Sanchez Alvarado A. Planarians. Curr Biol 2004;14(18):R737-738.

Sanchez Alvarado A. Regeneration and the need for simpler model organisms. Philos Trans R Soc Lond B Biol Sci 2004;359(1445):759-763.

Sanchez Alvarado A. Planarian regeneration: its end is its beginning. Cell 2006;124(2):241-245.

Sanchez Alvarado A, Kang H. Multicellularity, stem cells, and the neoblasts of the planarian Schmidtea mediterranea. Exp Cell Res 2005;306(2):299-308.

Sanchez Alvarado A, Newmark PA. The use of planarians to dissect the molecular basis of metazoan regeneration. Wound Repair Regen 1998;6(4):413-420.

Sanchez Alvarado A, Newmark PA. Double-stranded RNA specifically disrupts gene expression during planarian regeneration. Proc Natl Acad Sci U S A 1999;96(9):5049-5054.

Sanchez Alvarado A, Newmark PA, Robb SM, Juste R. The Schmidtea mediterranea database as a molecular resource for studying platyhelminthes, stem cells and regeneration. Development 2002;129(24):5659-5665.

Sarnat HB, Netsky MG. The brain of the planarian as the ancestor of the human brain. Can J Neurol Sci 1985;12(4):296-302.

Sarnat HB, Netsky MG. When does a ganglion become a brain? Evolutionary origin of the central nervous system. Semin Pediatr Neurol 2002;9(4):240-253.

Sato K, Sugita T, Kobayashi K, Fujita K, Fujii T, Matsumoto Y, Mikami T, Nishizuka N, Nishizuka S, Shojima K, Suda M, Takahashi G, Himeno H, Muto A, Ishida S. Localization of mitochondrial ribosomal RNA on the chromatoid bodies of marine planarian polyclad embryos. Dev Growth Differ 2001;43(2):107-114.

Sato Y, Kobayashi K, Matsumoto M, Hoshi M, Negishi S. Comparative study of eye defective worm 'menashi' and regenerating wild-type in planarian, Dugesia ryukyuensis. Pigment Cell Res 2005;18(2):86-91.

Sauzin-Monnot MJ. [Action of broken fractions of regeneration blastema on the synthetic activity of posterior fragments or Dendrocelum lacteum cut behind the pharynx]. C R Acad Sci Hebd Seances Acad Sci D 1976;282(21):1885-1888.

Sawada T, Oofusa K, Yoshizato K. Characterization of a collagenolytic enzyme released from wounded planarians Dugesia japonica. Wound Repair Regen 1999;7(6):458-466.

Schaeffer DJ. Planarians as a model system for in vivo teratogenesis studies. Qual Assur 1993;2(3):265-318.

Schaeffer DJ. Planarians as a model system for in vivo tumorigenesis studies. Ecotoxicol Environ Saf 1993;25(1):1-18.

Schaeffer DJ, Goehner M, Grebe E, Hansen LG, Hankenson K, Herricks EE, Matheus G, Miz A, Reddy R, Trommater K. Evaluation of the "reference toxicant" addition procedure for testing the toxicity of environmental samples. Bull Environ Contam Toxicol 1991;47(4):540-546.

Schaeffer DJ, Tehseen WM, Johnson LR, McLaughlin GL, Hassan AS, Reynolds HA, Hansen LG. Cocarcinogenesis between cadmium and Aroclor 1254 in planarians is enhanced by inhibition of glutathione synthesis. Qual Assur 1991;1(1):31-41.

Schiffmann Y. An hypothesis: phosphorylation fields as the source of positional information and cell differentiation--(cAMP, ATP) as the universal morphogenetic Turing couple. Prog Biophys Mol Biol 1991;56(2):79-105.

Schilt J, Richoux JP, Dubois MP. [Demonstration of peptides immunologically related SRIF, neurophysins and angiotensin in the planarian Dugesia lugubris (Platyhelminths,Turbellaria) (author's transl)]. J Physiol (Paris) 1981;77(8):977-978.

Schilt J, Richoux JP, Dubois MP. Demonstration of peptides immunologically related to vertebrate neurohormones in Dugesia lugubris (Turbellaria, Tricladida). Gen Comp Endocrinol 1981;43(3):331-335.

Schwartz E. [The rheotropism of Planaria lugubris O. Sch.; quantitative investigation.]. C R Seances Soc Biol Fil 1952;146(9-10):768-771.

Schwartz E. [Quantitative study of chemotropisms of Planaria [= Dugesia] lugubris O. Sch.; effects of a base, an acid and a mineral salt.]. C R Seances Soc Biol Fil 1953;147(9-10):879-882.

Sedensky MM, Hudson SJ, Everson B, Morgan PG. Identification of a mariner-like repetitive sequence in C. elegans. Nucleic Acids Res 1994;22(9):1719-1723.

Sengel C. [Culture in vitro of regeneration blastemas of Planaria.]. C R Hebd Seances Acad Sci 1959;249:2854-2856.

Sengel C. [In vitro culture of regeneration blastemata in Planaria.]. J Embryol Exp Morphol 1960;8:468-476.

Shafer JN, Corman CD. Response of planaria to shock. J Comp Physiol Psychol 1963;56:601-603.

Shagin DA, Barsova EV, Bogdanova E, Britanova OV, Gurskaya N, Lukyanov KA, Matz MV, Punkova NI, Usman NY, Kopantzev EP, Salo E, Lukyanov SA. Identification and characterization of a new family of C-type lectin-like genes from planaria Girardia tigrina. Glycobiology 2002;12(8):463-472.

Sheiman IM, Kreshchenko ND, Sedel'nikov ZV, Groznyi AV. [Morphogenesis in planarians Dugesia tigrina]. Ontogenez 2004;35(4):285-290.

Sheiman IM, Sakharova N, Tiras Kh P, Shkutin MF, Isaeva VV. [Regulation of asexual reproduction in the planarian Dugesia tigrina]. Ontogenez 2003;34(1):43-49.

Sheiman IM, Sedel'nikov ZV, Kreshchenko ND. [Resources of regeneration in planarians]. Ontogenez 2006;37(1):27-31.

Sheiman IM, Sedel'nikov ZV, Shkutin MF, Kreshchenko ND. [Asexual reproduction of planarians: metric studies]. Ontogenez 2006;37(2):130-135.

Sheiman IM, Tiras Kh P, Aptikaeva GF, Pavulsone SA. [Extracts of regenerating Planarians regulate proliferation of vertebrate cells]. Ontogenez 1987;18(5):546-550.

Sheiman IM, Tiras Kh P, Balobanova EF. [The morphogenetic function of neuropeptides]. Fiziol Zh SSSR Im I M Sechenova 1989;75(5):619-626.

Sheiman IM, Zubina EV, Kreshchenko ND. [Regulation of appetitive behavior in planaria Dugesia (Girardia) tigrina]. Zh Evol Biokhim Fiziol 2002;38(4):322-325.

Shibata N, Umesono Y, Orii H, Sakurai T, Watanabe K, Agata K. Expression of vasa(vas)-related genes in germline cells and totipotent somatic stem cells of planarians. Dev Biol 1999;206(1):73-87.

Shinkman PG, Vernon LM. An Apparatus for Injecting Planarians. Percept Mot Skills 1965;20:726-728.

Silveira M, Corinna A. Fine structural observations on the protonephridium of the terrestrial triclad Geoplana pasipha. Cell Tissue Res 1976;168(4):455-463.

Simoncelli F, Sorbolini S, Fagotti A, Di Rosa I, Porceddu A, Pascolini R. Molecular characterization and expression of a divergent alpha-tubulin in planarian Schmidtea polychroa. Biochim Biophys Acta 2003;1629(1-3):26-33.

Slack JM. The source of cells for regeneration. Nature 1980;286(5775):760.

Slack JM. Regeneration research today. Dev Dyn 2003;226(2):162-166.

Steele VE, Lange CS. Effects of irradiation on stem cell response to differentiation inhibitors in the planarian Dugesia etrusca. Radiat Res 1976;67(1):21-29.

Steele VE, Lange CS. Characterization of an organ-specific differentiator substance in the planarian Dugesia etrusca. J Embryol Exp Morphol 1977;37(1):159-172.

Stephan-Dubois F, Bautz A. [Histological and cytological study of neoblasts and fixed parenchymal cells in Dendrocoelum lacteum planarians deprived of their anterior region]. Arch Anat Microsc Morphol Exp 1975;64(1):75-89.

Stornaiuolo A, Bayascas JR, Salo E, Boncinelli E. A homeobox gene of the orthodenticle family is involved in antero-posterior patterning of regenerating planarians. Int J Dev Biol 1998;42(8):1153-1158.

Stringer CE. The Means of Locomotion in Planarians. Proc Natl Acad Sci U S A 1917;3(12):691-692.

Stuart GW, Zhu Z, Sampath K, King MW. POU-domain sequences from the flatworm Dugesia tigrina. Gene 1995;161(2):299-300.

Sumiyama K, Washio-Watanabe K, Saitou N, Hayakawa T, Ueda S. Class III POU genes: generation of homopolymeric amino acid repeats under GC pressure in mammals. J Mol Evol 1996;43(3):170-178.

Tarabykin VS, Lukyanov KA, Potapov VK, Lukyanov SA. Detection of planarian Antennapedia-like homeobox genes expressed during regeneration. Gene 1995;158(2):197-202.

Tazaki A, Gaudieri S, Ikeo K, Gojobori T, Watanabe K, Agata K. Neural network in planarian revealed by an antibody against planarian synaptotagmin homologue. Biochem Biophys Res Commun 1999;260(2):426-432.

Tazaki A, Kato K, Orii H, Agata K, Watanabe K. The body margin of the planarian Dugesia japonica: characterization by the expression of an intermediate filament gene. Dev Genes Evol 2002;212(8):365-373.

Tehseen WM, Hansen LG, Schaeffer DJ, Reynolds HA. A scientific basis for proposed quality assurance of a new screening method for tumor-like growths in the planarian, Dugesia dorotocephala. Qual Assur 1992;1(3):217-229.

Tei S. [Preliminary findings on some effects of a punctiform magnetic field on regeneration in Dugesia lugubris]. Boll Soc Ital Biol Sper 1975;51(6):324-328.

Tiras Kh P. [Acetylcholinesterase activity in the nervous system of normal planaria and during regeneration]. Ontogenez 1978;9(3):262-268.

Tiras Kh P, Aslanidi KB. [Device for graphic recording of planarian behavior]. Zh Vyssh Nerv Deiat Im I P Pavlova 1981;31(4):874-877.

Tiras KP, Sakharova NY, Sheiman IM. [Acetylcholinesterase activity in the nervous system of some Triclades (Planaria)]. Zh Evol Biokhim Fiziol 1975;11(4):427-429.

Tushmalova NA. [Conditioned reflexes in Baikal planaria, Podoplana olivacea K. following an injection of ribonuclease]. Zh Vyssh Nerv Deiat Im I P Pavlova 1967;17(2):359-361.

Tushmalova NA. [Phylogenetic age of conditioned reflex memory (review)]. Zh Evol Biokhim Fiziol 1980;16(5):506-515.

Tutel'ian VA, Vasil'ev AV. [Lysosomal proteolytic system of eukaryotes in the process of phylogenetic development]. Zh Evol Biokhim Fiziol 1982;18(2):113-118.

Tyder TA, Bowen ID. Letter to the editor: A method for the fine structural localization of acid phosphatase activity using rho-nitrophenyl phosphate as substrate. J Histochem Cytochem 1975;23(3):235-237.

Ugazio G, Burdino E, Crespi M, Barbero N, Garizio M, Arru G, Congiu AM. [Eco-toxicological study conducted with a battery of biological and phytological tests on sediments carried out on a series of 24 tributaries of the Po in 1994 and 1995]. G Ital Med Lav Ergon 1997;19(1):10-16.

Umeda S, Stagliano GW, Borenstein MR, Raffa RB. A reverse-phase HPLC and fluorescence detection method for measurement of 5-hydroxytryptamine (serotonin) in Planaria. J Pharmacol Toxicol Methods 2005;51(1):73-76.

Umeda S, Stagliano GW, Raffa RB. Cocaine and kappa-opioid withdrawal in Planaria blocked by D-, but not L-, glucose. Brain Res 2004;1018(2):181-185.

Umesono Y, Watanabe K, Agata K. A planarian orthopedia homolog is specifically expressed in the branch region of both the mature and regenerating brain. Dev Growth Differ 1997;39(6):723-727.

Umesono Y, Watanabe K, Agata K. Distinct structural domains in the planarian brain defined by the expression of evolutionarily conserved homeobox genes. Dev Genes Evol 1999;209(1):31-39.

Vagnetti D, Farnesi RM, Tei S, Marinelli M. [Ultrastructural data on some functional aspects of the copulatory pouch of Dugesia lugubris s.l]. Boll Soc Ital Biol Sper 1975;51(9-10):582-587.

van der Linden AG. Spermatogenesis, spermiogenesis and sperm structure in Crenobia (Planaria) alpina (Dana). Z Zellforsch Mikrosk Anat 1969;97(4):549-563.

Van der Linden AG. Chromosome number and ploidy-level in a Dutch population of Crenobia alpina Dana (Planaria). Genetica 1969;40(1):61-64.

Vandeventer JM, Ratner SC. Variables Affecting the Frequency of Response of Planaria to Light. J Comp Physiol Psychol 1964;57:407-411.

Venturini G, Carolei A, Palladini G, Margotta V, Lauro MG. Radioimmunological and immunocytochemical demonstration of Met-enkephalin in planaria. Comp Biochem Physiol C 1983;74(1):23-25.

Venturini G, Stocchi F, Margotta V, Ruggieri S, Bravi D, Bellantuono P, Palladini G. A pharmacological study of dopaminergic receptors in planaria. Neuropharmacology 1989;28(12):1377-1382.

Viaud G. [Galvanotropic reactions of isolated pharynx of Planaria lugubris O. Sch.]. C R Seances Soc Biol Fil 1950;144(17-18):1203-1206.

Viaud G. [Anisotropy on galvanic excitation and electric anisotropy of segments of pharynx isolated from Planaria (Dugesia) lugubris O. Schm.]. C R Seances Soc Biol Fil 1952;146(17-18):1382-1384.

Viaud G. [Electrical anisotropy and variations of conductance as a function of section in Planaria lugubris O. Sch.]. C R Seances Soc Biol Fil 1952;146(5-6):489-492.

Viaud G. [Galvanotropism, excitatory anisotropy and electrical anisotropy in a planarian (Planaria Dugesia lugubris O. Schm.)]. J Physiol (Paris) 1952;44(2):343-345.

Viaud G. [Quantitative study of the electromotor force of opposition produced by planaria (Planaria-Dugesia lugubris O. Schm.) in response to an electric excitation due to continuous current.]. C R Seances Soc Biol Fil 1954;148(23-24):2068-2071.

Viaud G. [Experimental study of galvanotropism of Planaria: cathodic direction of the tropism and electric anisotropy of Planaria lugubris.]. Annee Psychol 1954;54(1):1-33.

Viaud G. [Inversion of the direction of galvanotropism of Planaria (=Dugesia) lugubris O. Schm. by the action of polarized visible and ultraviolet radiations.]. C R Seances Soc Biol Fil 1955;149(23-24):2221-2224.

Viaud G, Medioni J. [Anisotropy of excitation and variations of the thresholds of reaction to galvanic current in relation to size in Planaria lugubris, O. Sch..]. C R Seances Soc Biol Fil 1951;145(15-16):1228-1231.

Villar D, Gonzalez M, Gualda MJ, Schaeffer DJ. Effects of organophosphorus insecticides on Dugesia tigrina: cholinesterase activity and head regeneration. Bull Environ Contam Toxicol 1994;52(2):319-324.

Villar D, Li MH, Schaeffer DJ. Toxicity of organophosphorus pesticides to Dugesia doroto-cephala. Bull Environ Contam Toxicol 1993;51(1):80-87.

Villar D, Schaeffer DJ. Morphogenetic action of neurotransmitters on regenerating planarians--a review. Biomed Environ Sci 1993;6(4):327-347.

Vispo M, Cebria F, Bueno D, Carranza S, Newmark P, Romero R. Regionalisation along the anteroposterior axis of the freshwater planarian Dugesia(Girardia)tigrina by TCEN49 protein. Int J Dev Biol 1996;Suppl 1:209S-210S.

Vladimirova IG. Respiration during wound regeneration and healing in the planarian worm Dendrocoelum lacteum. Biol Bull Acad Sci USSR 1980;7(5):435-439.

Vykhrestiuk NP, Nikitenko TB, Iarygina GV. [Lipids of free-living Penecurva sibirica turbellaria]. Zh Evol Biokhim Fiziol 1976;12(4):362-364.

Wago H. [Host defensive protein of planaria]. Tanpakushitsu Kakusan Koso 2001;46(4 Suppl):408-413.

Webb RA, Friedel T. Isolation of a neurosecretory substance which stimulates RNA synthesis in regenerating planarians. Experientia 1979;35(5):657-658.

Welsh JH, Williams LD. Monoamine-containing neurons in planaria. J Comp Neurol 1970;138(1):103-115.

Wilhelmi G. [Effect of antiphlogistic drugs on regenerative processes in Planaria gonocephala and Amblystoma mexicanum.]. Naunyn Schmiedebergs Arch Exp Pathol Pharmakol 1953;218(1-2):101-103.

Wilhelmi G. [The effect of phenothiazine preparations on the regeneration process of planaria and axolotl.]. Helv Physiol Pharmacol Acta 1955;13(2):C40-42.

Winsor L. Oblique illumination and interference contrast microscopy aids in the taxonomic histology of land planarians. Mikroskopie 1978;34(11-12):319-321.

Winsor L. Pseudoparasitism of domestic and native animals by geoplanid land planarians. Aust Vet J 1980;56(4):194-196.

Winsor L. Vomiting of land planarians (Turbellaria: Tricladida: Terricola) ingested by cats. Aust Vet J 1983;60(9):282-283.

Wolff E, Sengel P, Sengel C. [Is the caudal region of the Planaria capable of inducing regeneration of the pharynx.]. C R Hebd Seances Acad Sci 1958;246(11):1744-1746.

Yoshioka Y, Ose Y, Sato T. Correlation of the five test methods to assess chemical toxicity and relation to physical properties. Ecotoxicol Environ Saf 1986;12(1):15-21.

Yui R, Iwanaga T, Kuramoto H, Fujita T. Neuropeptide immunocytochemistry in protosto-
mian invertebrates, with special reference to insects and molluscs. Peptides 1985;6 Suppl
3:411-415.

Zamora-Veyl FB, Guedes HL, Giovanni-De-Simone S. Aspartic proteinase in Dugesia tigrina
(Girard) planaria. Z Naturforsch [C] 2002;57(5-6):541-547.

Zayas RM, Bold TD, Newmark PA. Spliced-leader trans-splicing in freshwater planarians. Mol
Biol Evol 2005;22(10):2048-2054.

Zayas RM, Hernandez A, Habermann B, Wang Y, Stary JM, Newmark PA. The planarian
Schmidtea mediterranea as a model for epigenetic germ cell specification: analysis of ESTs
from the hermaphroditic strain. Proc Natl Acad Sci U S A 2005;102(51):18491-18496.

Zghal F, Tekaya S. [The Turbellaria of the Gulf of Tunis. II. Studies on the regenerative
power in marine Planarians, Sabussowia dioica (Claparede 1863)]. Arch Inst Pasteur Tunis
1982;59(4):587-604.

Zhang HC, Chen GW, Li YC, Xu CS. [Optimization of reaction conditions for RAPD analysis
of freshwater planarians in China]. Shi Yan Sheng Wu Xue Bao 2004;37(4):333-336.

Ziller C. [The effect of repeated amputations on planarian regeneration in the presence of
the heat-stable toxin from Bacillus thuringiensis]. C R Acad Sci Hebd Seances Acad Sci D
1976;283(4):371-373.

Ziller C. [Effect of actinomycin D and cycloheximide on planarian regeneration after repeated
amputation]. C R Acad Sci Hebd Seances Acad Sci D 1976;283(3):251-254.

INDEX